CONSCIOUSNESS AND THE SEARCH FOR REALITY

CONSCIOUSNESS AND THE SEARCH FOR REALITY

Beyond the veil

Michele T. Knight

A Salon de Morte Publication

A Salon de Morte Publication

For information address:

Dr Michele T. Knight

salondemorte@gmail.com

Cover Photograph: Mi Serenata

By Robert Bryce, Studio Temenos © 2020

ISBN: 978-0-6459028-0-8 (paperback)

ISBN: 978-0-6459028-1-5 (kindle)

NATIONAL LIBRARY OF AUSTRALIA

A catalogue record for this book is available from the National Library of Australia

DEDICATION

Dedicated to the discontent and to those seeking new directions in their search for reality.

CONTENTS

Preface - The Seeds of Discontent. ix
Introduction - Integrative Studies 1
1 - Toward Eternity 21
2 - School Remembrances 43
3 - Know Thyself. 59
4 - Technical Terms 75
5 - Work Basics 89
6 - Faith 117
7 - What Am I? 131
8 - Intentional Effort 149
9 - A Word About Death 165
10 - A New Way. 173
Afterword - A Final Note. 185
Acknowledgments and Permissions 189
References 199

PREFACE

THE SEEDS OF DISCONTENT

Since childhood, discontent has been a life-long companion of mine, unfortunately. And for as long as I can remember accompanying that persistent internal and frustrating state was the feeling that there was more to life, but I didn't know what, and that something important, something integral was missing from mine, but I didn't know why. As if that wasn't enough, I was also constantly distracted by an uncontrollable internal driving urge to know and understand what and why, which sat inside me tapping away incessantly at my awareness.

Compounding this dilemma was another one; that it was also as a small child that I began to experience two co-existing or overlapping realities. I never talked about this with my parents or close family and to my then three-year old mind these realities were just different 'places' which I accepted existed because I experienced them. It was just the way things were, the way things are.

Writing about this now some fifty or so years afterward, I understand why the foundation for how I was to live in the world from that time on was laid down then and not later when I was an adult. It seems to me that children are innate, they feel and know things within themselves which have not yet been overtaken by the impressions of a noisy external life and a know-it-all ego and intellect. Such a state enables them to experience the world around them differently compared to when they become adults; I know I did.

Of the two realities I experienced, one was tangibly material because it was the world I lived in as a sentient, embodied human being. The second reality was characterised by non-material, other-worldly, or what I define as spiritual phenomena, and though these experiences populated my childhood and teenage years, I had no understanding of why they occurred. But because of them, and from a young age, I knew beyond a shadow of a doubt that something profoundly vast lay behind what my senses told me was 'reality'. As I matured into young adulthood, I recalled always wanting to have a sense of permanence in at least one of these realities, but only ever being beset by shadows from both.

When I was twenty years of age that changed because I encountered a body of knowledge, some would say a doctrine, which led me to become a student at an esoteric Christian School, and how I lived in the world and how I understood myself and life was never the same again. It was never the same again because I realised that while I believed I was alive, I was the recipient of life, and while I knew of the existence of another kind of reality, an other-worldly or spiritual reality, I had no understanding of the significance of that reality. I knew physical death occurred during our existence, but I didn't know what I was living for or why I had been born, or what would happen to me on the other side of that death. And I wanted to know. I needed to know.

This book, which I was asked to write, chronicles my engagement with those fundamental existential questions and my journey with the School and the education I received which provided so many answers and explanations for them. When we hear the term 'school', we tend to equate it, especially in the West with mainstream teaching and education institutions, but this School was different. It was different because it taught a different type of knowledge and because it had a different type of purpose (I write at length about the School in the second chapter).

Many memoirs and autobiographies of pupils who attended similar Schools in other countries have been published, and while such publications might be loosely interpreted as a type of hagiography whereby the author inspires an ongoing closeness with the teacher after their death or even a reverence for them through their writing, this book is not of that genre. While this book is semi-autobiographical it was never my intention to memorialise or revere the School's teacher in any way, and to do so would be counter to the book's purpose. Furthermore, the book was written while the teacher was still living, it's simply been a long time in coming to print.

Human beings are complex, multi-layered, and multi-levelled structures. As cosmic instruments, they have the capability of being able to transform themselves and to maximise their spiritual potentials to metamorphose into a living, breathing, conscious soul. When this occurs, the soul is then able to fulfil its cosmic duties and can assume its rightful place in the spiritual universe after physical death. This is a truth largely unknown in post-modern society, yet in esoteric schools one that is not only taught and known among their pupils, it is a living reality.

As a result of being a student in this School, it now feels that the past, the present and the future continually merge and blend into a unique and dynamic state of being which I experience as 'now'. In this state of being it appears that all events coalesce or merge into an ongoing experience of life. There is no 'past' or 'future', there are only moments or events which appear to be measurable fragments of time yet framing those moments or perhaps giving them shape and structure, is something profoundly spiritually abiding. From this abiding there comes an unveiling, tentative at first until gradually a perpetual revealing of the world as it turns on the axis of the universe becomes visible. And at times, barely discernible, there's a pulse, a rhythm almost, that seems to beat or throb in time with something beyond the senses.

Michele T. Knight, Sydney, Australia, December 10, 2023

INTRODUCTION

INTEGRATIVE STUDIES

> Esoteric knowledge is the science of man's relation with God and the universe. Its transmission requires an engagement with others – so-called "schools" – because a certain energy can only be produced in conditions where people work together. Schools may differ in their knowledge and their approach – their way – but they have the same aim in common: to see reality. The knowledge is passed on theoretically and through direct experience, that is by living a drama which follows the particular way of the school. This creates a relation, the link without which it would not be possible to live in two worlds of different levels at the same time.
>
> Jeanne de Salzmann, *The Reality of Being: The Fourth Way of Gurdjieff*, p.2, 2010.

In February 1985 I became a student in an esoteric Christian School which I attended for the next fourteen years (the term 'School', 'esoteric Christian School' and 'esoteric School' are used interchangeably throughout). It was Summer and I was twenty years old, and though I didn't know it at the time, joining the School heralded the commencement of what is termed my Second Education and my introduction to the principles and practices of The Fourth Way or what is generically known as, the Work. Second Education is a technical term which denotes a specific kind of knowledge and activity orientated toward self-transformation, spiritual regeneration, and growth of being.

In the pages and chapters which follow, I peel away the layers of what was taught in the School and share the intersection of that education with the lived experience of a diverse range of spiritual or other-worldly phenomena. (The terms 'spiritual' and 'other-worldly' are used interchangeably throughout and are analogous with peak-experiences, mystical experiences, anomalous experiences, and transpersonal experiences.)

What may be unfamiliar terms and phrases populate the book, so I have included an explanatory chapter, *Technical Terms,* which provides some context for readers who may be new to them. I recall being confused when I first encountered them myself, and it took me some time to understand exactly what they meant; it was a language unlike any I knew. Because of this, I encourage the reader not to be dissuaded in their own attempts to understand the terms when encountered.

One term in particular 'consciousness', is included in the title of this book. The reason for this is that attaining consciousness is one of the driving aims of the Work and of work-on-oneself, the means of which was taught and frequently discussed in the School. It was taught because of its important relationship with the spiritual pattern of reality. Consciousness is spiritual however it can often be confused or mistaken with *awareness.* It is not awareness; it is something quite different.

At the time of joining the School I was considered by other more mature students who had been attending for many years to be quite young, however when someone enters the Work they do so at the time that is right for them; chronological age has no relevance whatsoever. Located in Sydney, Australia, the School was open to all and although students came and went there was always a core group who regularly attended. Students were sometimes absent for varying periods, often because of work and family commitments or health issues, but in reflecting on that time now, I feel the separation afforded them the opportunity to digest the impressions of the Work and what it was we were being taught.

I was no different to the many who attended in that I combined being a student with living a busy external or daily life. I was working in the advertising industry, was in a relationship, was completing my first degree at The University of Sydney and I had an active social life with lots of friends which I juggled with an array of interests and hobbies. I didn't suddenly stop living my life, or retire from it, nor did I cut myself off from family and friends or stop participating in the things I loved doing because I attended the School. What I did do, was simply incorporate being a student at the School into my daily life, in the same way I did with my university studies.

When I first met the school's teacher, Dr Philip W. Groves as a shy young adult, he told me that the spiritual experiences I had intuitively recorded in my diaries and journals throughout the years would provide the material for books that would one day be written. "I see books on shelves" he had said to me, as if that explained everything (*Dr Philip W. Groves, personal communication, 2 March 1985*). When sharing one of my spiritual experiences with him some months after that initial conversation, he explained the reason for their occurrence in the following words. "You have a very great spiritual sensitivity. It seems to me that you are being prepared for some sort of useful work in life that concerns the awakening of men and women to their spiritual potentials, hence you must pass through a wide range of spiritual experiences so that you can know the truth of them at first hand" (*Dr Philip W. Groves, personal communication, 14 December 1985*). What on earth did *that* mean?

It was the Summer of 1998 and some thirteen years after I had first met Dr Groves and joined the School, that he charged me with the task of writing this book. We were talking together one Saturday morning, and all these years later I can still see the bright sunlight shining through the lounge-room bay windows as if it were only yesterday. I was sitting on one of the overstuffed couches which always crowded the lecture room (and from which I could never extricate myself with any semblance of dignity in all the years I sat on them), propped up by similarly overstuffed cushions strategically positioned behind my back.

He looked at me rather directly during the conversation and said unexpectedly, "I want you to put all of the Work into a book." My ears began to ring, and I noticed a pounding sensation in my chest. Was that perspiration forming on the palms of my hands? I had written my first 'book' at the age of eleven, a dashing account of hidden treasure on a deserted island and since that time though I had kept diaries and journals, and had written poems and prose, I had never attempted a book, much less one claiming to be a vehicle for the Work. "I can't do that" I responded incredulously. "Yes, you can" he said. Thinking him completely mad, I responded, "No, no I can't." He looked at me, smiled, and said, "Yes, you can."

My doubts remained and so I persisted. I wasn't learned or intellectual or particularly clever, and I certainly wasn't like other writers who had an extensive and formidable knowledgebase of the Work and its doctrine. I hadn't run Work groups. I hadn't even participated in a Zikr (and to this day still never have), and I had trouble remembering to remember myself; I considered myself in every way possible not at all qualified or suitable for the task. I even questioned why he didn't do it and couldn't help but laugh when he told me with great earnestness that he had written a manuscript about the Work, but its quality had been such that at the time he considered his only option was to throw it away, which he did. Despite my protestations he was adamant, saying that he wanted the Work to be experienced from a loving and affectionate perspective rather than from one strictly intellectual.

He told me then that nobody owned or had ownership of the Work, that it was available to all, and that I had free licence to write about it and to quote and reference him as frequently and as necessary as I saw fit to do. I could decide on the format and content of the book and could draw directly from any of the verbal or written material he had provided during lectures to students at the School, and from any of his public lectures and published written material. I was also permitted to draw from our private conversations, and from my personal journal notes referring to the School and its teaching which I had written and compiled during my attendance. He would review the chapters as I wrote them and would provide any necessary comments or feedback.

In addition to including select events which had occurred during my adult life, he particularly wanted accounts of my childhood spiritual experiences included, telling me that the book should be semi-autobiographical to demonstrate the Work in daily life by way of lived experience. Reflecting on his words some days afterward, I felt at the time that perhaps he wanted these accounts included because while for me they provide a context of *being*, which is relevant to work-on-oneself and to the Work in general, they also demonstrate something else; a relationship with spiritual reality which is attainable by all.

A product of esoteric Schools himself, Dr Groves taught in his School the doctrine he termed *Beyond the Fourth Way*, the principles and practices of which are set down in the published text of the same name (The Triam Press, 1996). At the time I met him, he had already been a teacher of the Work for over fifty years, teaching not only the principles of The Fourth Way but also introducing and teaching, in Australia, the principles and practice of Beyond the Fourth Way. There were numerous occasions during lectures to students and in his published material whereby he explained the difference between the Fourth Way and Beyond the Fourth Way, which I have contextualized as follows.

The Fourth Way which comprises the core of the Work, is principally a system and methodology of self-study. Its aim is to reorientate the individual from solely looking outward to the world, to looking inward to bring about self-change. It is utilised by the individual to awaken from the universal psychospiritual 'sleep' of mankind (the terms 'Man' and 'Mankind' are used interchangeably throughout and represent the collective non-gendered human race) otherwise known as the sleep-state. It is an ancient name given to a particular line of esoteric teaching which encourages people to attain consciousness, a process which occurs alongside discrete stages of being-growth.

These different stages of growth and development of consciousness are illustrated by the Ray of Creation, which sets out in ascending order different levels of reality represented by worlds, which in turn reflect the stages of consciousness. But this is suggestive of a somewhat finite and linear trajectory in that when consciousness is attained the pinnacle of work-on-oneself has been reached, which is not the case. Since its introduction to the West, the Fourth Way has planted the knowledge and understanding of self-transformation in countless minds, however in the fifty years of his teaching the Work what Dr Groves observed and shared was that students wanted to know where the Fourth Way was leading them, and where they were going in the spiritual universe.

The Fourth Way was introduced to the West by Greek-Armenian Georges Ivanovitch Gurdjieff (1866-1949). In 1915 the inception of Gurdjieff's system, which he referred to as esoteric Christianity, took root, commencing with a study group in St Petersburg, Moscow. Various locations for the group followed in the intervening years and then in 1922, Gurdjieff established the second Institute for the Harmonious Development of Man, at the Chateau du Prieuré in Fontainebleau-Avon, not far from the city of Paris. As the result of a car crash in 1924 however, Gurdjieff closed his institute and spent the latter part of his life writing and occasionally traveling abroad.

Since then, and in addition to Gurdjieff, the Fourth Way has been promulgated by such exponents as Peter D. Ouspensky (1878-1947), Alfred R. Orage (1873-1934), John G. Bennett (1897-1974), Maurice Nicoll (1884-1953), Beryl Pogson (1895-1967), Margaret C. Anderson (1908-1973), Kathryn Hulme (1900-1981), and Dr Philip W. Groves (1920-1999). This is not an exhaustive list, and now with the advent of the world-wide web and social media, many others have since set up and run groups and courses, manage associations, and offer programs which teach the Fourth Way.

Jeanne de Salzmann (1889-1990), close friend and core group member of Gurdjieff's Work groups, notes that in his later years Gurdjieff considered the study of his original system of ideas a preliminary stage of the work toward consciousness. He also considered questions regarding the ideas theoretical, and instead presented his teaching in terms of what was considered a "direct perception of reality" (De Salzmann, 2010, p. 296). The Fourth Way, which is the core method of the Work is carried out during daily life and while it provides critical instruction for how to awaken from the sleep-state, what then? What comes next? The consideration is twofold; what are we waking up for, and what lies beyond the process of awakening?

When Gurdjieff wrote his *All and Everything* series, (ARKANA,

1950), although he indicated that there were other planes and other levels of attainment and reality, he did not specifically write about their function or the processes of how to reach them (*Author's lecture notes, 12 December 1998*). Apart from what might have constituted his personal belief, knowledge and understanding about such matters, and although he may have hinted at it or perhaps discussed it in private, it appears that Gurdjieff did not elaborate publicly about what lies beyond the process of awakening. It was in response to this that Beyond the Fourth Way came into being, it complements the crucial foundation provided by the Fourth Way taught by Gurdjieff because it teaches people how to reach those other planes and levels.

Although in private the Fourth Way teachers, writers and students may have maintained their own loving relationships with the Divine and Higher Life little appears to have been written regarding how to combine the intellectual knowledge of the Work with the process of establishing a loving relationship with God as well, nor of the reason or importance for doing so. This is what Dr Groves wanted me to make known; how that can occur and what happens when it does. Terms such as 'God', 'the Divine', and 'Higher Life' are used interchangeably throughout this book and are synonymous with each other. 'Higher Life' for example is an encompassing term which acknowledges the *workings* of God; the creation and maintenance of the material and spiritual universe, and all and everything existing within those universes.

The lived experience of my life has evidenced that I come from a spiritual source, that I need to create my own spiritual destiny and that this is not possible without help from Higher Life and a loving relationship with God. The foundation for what I didn't know would become life-long spiritual effort and work was laid in my childhood. This was a time during which I not only constantly felt that something deep and abiding lay beyond physical reality, but that it sustained me in a fundamental way. Although I didn't understand what that was or how it did so, I felt the certainty of it within myself and intuitively knew it to be true. Even then and as young as I was, I was able to observe two very distinct parts of myself, one which thought and one which felt.

God is the axis of the Work which manifests through the vehicles of the Fourth Way and Beyond the Fourth Way; this is an important principle. We do need to wake up, that is essential, but we need to know what we're waking up *for* as we do *why*. We also need to understand how to awaken, as we do what it *is* that awakens us. And this is where our relationship with the Divine plays a pivotal role.

> Beyond the Fourth Way is not an attempt to undermine or supersede the Fourth Way. That basic teaching is monumental and can never be replaced. In Beyond the Fourth Way, we seek to know the spiritual foundations of reality and the eternal prospects which lie before our soul. (Groves, 1998f, p. 13).

Beyond the Fourth Way tells us that there is a spiritual beginning and a spiritual finality, and while the Fourth Way pervades the body of the Work, "without Beyond the Fourth Way it is useless" (*Dr Philip W. Groves, author's lecture notes, 16 December 1998*). What Dr Groves was trying to teach students, was that the Second Education referred to in esoteric writings can combine the principles of both the Fourth Way *and* Beyond the Fourth Way. What was always emphasised in the School was that a relationship with God is central to the Work and to work-on-oneself.

There is a relationship between these principles and with the establishment and introduction into mankind of the 'New Church', which Swedish scientist, philosopher, theologian, and mystic Emanuel Swedenborg (1688-1772), wrote extensively about. In several texts devoted to the subject (see *The New Jerusalem and its Heavenly Doctrine, The Cannons of the New Church, A Brief Exposition of the Doctrine of the New Church, Four Doctrines of the New Jerusalem*) Swedenborg reiterates throughout that the term 'The New Church' signifies the New Jerusalem written of in Revelation, Chapters xxi and xxii. According to Swedenborg, the church is "the foundation of heaven" (1758/1993, p. 135), and regarding the Work and work-on-oneself, it is important to understand the context in which 'church' is used, what it means, and what the implications are.

As I've mentioned, the language of the Work abounds with what are known as technical terms, one of which is *Church.* These terms are correspondences, and contrary to its contemporary literal understanding in the West, in the Work the term church corresponds to an orientation of mind.

> Church is a very ancient term, which fundamentally indicates a functional relationship between God and man. It is something which has an infinite spread. God doesn't confine Himself to people who pray in a certain way. God has an infinite spread because man is the crowning glory. We have a soul, this is our destiny, and we must make this destiny come true for ourselves. The relationship between the Divine and every human being is called the Church. (*Author's lecture notes, 16 December 1998*).

When we work on ourselves, we enter into a partnership with the Divine and inwardly become the representation of a church. This occurs because we pave the way for the reception of specific influences, which come from God, into ourselves. In time we come to understand the reality of our cosmic situation as we do our complete and total reliance on the Divine who affects all things, including our self-cleansing and transformation.

The Work not only seeks to restructure the mind, changing it from a sleeping to a wakeful state, it strives to bring the heart into a state of loving remembrance of God. When these are active within us, we are deemed to be a church. The mind is an individual's most important function because this is the part that looks up or down. It is the mind that is the instrument of growth and transformation. The destiny of the soul is wrapped up in the mind. People often say that they 'have a soul', this is incorrect, they should say, "I am a soul."

Awakening from the sleep-state is the primary task of work-on-oneself, and while the Work teaches the laws of world-building and world-maintenance, it is important to understand (from a Work perspective) that we're born into this world to transform our being from its bundle of potentials into actualities, and how we achieve that is by establishing a functional and loving relationship with God. We determine our eternal spiritual destiny in accordance with how we live our lives as embodied beings, and it is due time that a book be written discussing the group work of an esoteric Christian School which taught the spiritual reality of the soul's eternal prospects.

When Dr Groves wanted me to 'put all of the Work into a book', he specifically asked that I include my own private spiritual experiences, because they intersect with the Work and evidence Higher Life in action. During the first conversation I had with him shortly after we met, he had explained that my spiritual experiences and subsequent writing and intuitive recording of them throughout my life were addendums to those of Emanuel Swedenborg. When I asked him why such things occurred, he replied in the matter-of-fact manner I became accustomed to during the years I knew him, "These are part of your spiritual education and as they educate you, so they will educate others" (*Dr Philip W. Groves, personal communication, 2 March 1985*). As if that made any sense at all to me. It was so simple for him yet so tortuously complicated for me. What did it all *mean*?

I understand now, many years after those early conversations with him, that work-on-oneself heralds a journey of intimate engagement between the individual and the Divine. It is not only an intensely personal and profoundly transformative journey unique to the individual, but one which has far-reaching implications for their eternal existence. And with all esoteric literature pertaining to such, this book will also have its own unique identity and journey in society and in the hands of whoever reads it. It will not be more than or less than those written by others who too have discoursed on the growth and evolution of the human soul and the perfecting of the mind and spirit. It will sit side-by-side with other books yet be uniquely apart from them as well.

It is not without great trepidation that in the following pages I reveal intimate and very private aspects of my personal life, and I haven't done so lightly. It is my hope that the reader will appreciate them for what they are; life events which serve to educate and inform. While I acknowledge the difficulty of understanding the experiences of others without first perhaps having a frame of reference for them ourselves, everything I relate in the following pages is a truthful account of what has taken place in my life. These, as I perceive them, spiritually crafted and intentional events include such things as higher essence experiences, out-of-body experiences, near-death experiences, mystical experiences, dreams that were more than a dream, and direct interactions with spiritual or disembodied beings. The term 'disembodied' represents someone who has died and who now lives as a spiritual being in the spiritual universe. The person still exists, but in different form.

To the contemporary Western mind which is heavily influenced by the empirical sciences and captured fast by bodily senses, this may sound fantastical and I appreciate the challenge which faces the reader; it can be difficult believing something if we haven't experienced the same ourselves. For example, a higher essence experience occurs when there is a sensed movement from external or outer things to interior or internal things. This isn't a movement in the physical sense, but rather a change of internal state. How do we understand what these words mean especially if we've never experienced something like this? Herein lays the challenge.

I first began to experience these types of events in childhood and while I struggled to understand them then, I think I know why they commenced at that time rather than during my adult life; the unruly outer self could have been obstructive and possibly dismissive of them. Higher essence experiences provide us with a spiritual view of reality from the Divine perspective. They can be deeply profound and very powerful because they shock or jolt us into momentarily experiencing a different type of reality from that of the day-to-day to which we are accustomed.

I know that many of the personal experiences recounted in this book may be met with various judgements such as cynicism, disbelief, scorn, and ridicule. I speak to those judgements by standing by the words I write, and while I acknowledge that making public such private accounts may draw unwanted attention and possible derision, I am encouraged by the following words of Swedenborg.

> I am well aware that many will say that no one can possibly speak with spirits and angels so long as he lives in the body; and many will say that it is all fancy, others that I relate such things in order to gain credence, and others will make other objections. But by

> all this I am not deterred, for I have seen, I have heard, I have felt. (1751-1752/1984, AC 68, pp. 39-40).

While Beyond the Fourth Way enables us to "know the spiritual foundations of reality" (Groves, 1998f, p. 3), it is not by chance that the origin and nature of the universe has been the subject of, at times, a polarising discourse throughout past and present ages.

In the 16th Century for example, the Irish scholar Bishop James Ussher (1581-1656), a Calvinist prelate in the Church of Ireland, held that the universe came into being on 22nd October, in the year 4004 BC. In the 21st Century, scientists, cosmologists and astrophysicists, who once strictly adhered to the Big Bang theory which asserts that the universe came into existence approximately 13.7 billion years ago, or thereabouts, now acknowledge that the Big Bang theoretical construct is one of a number of cosmological models, which according to South African Cosmologist Professor George F. R. Ellis (1939-), explain various observations about the origin and nature of the universe.

My voice echoes others past and present, in arguing that it isn't only the contemporary sciences' which need to question the principles of physicalism it is all of us who need to do so, and I have long discoursed privately and publicly, as others have done before me and no doubt will continue to do so after me, that greater acknowledgement be given to a paradigm encompassing a spiritual foundation of reality. We need to understand that we live in a material universe which is spiritual in origin.

> It is true that physics and chemistry underlie our existence and functioning as human beings, but that does not mean we are "nothing but" atoms, molecules, chemicals, or whatever. That phrase always hides an attempt to deny the true complexity and auton-

> omous existence of vibrant living beings. We are much, much more than implied by hard reductionists and their favourite phrase "nothing but" ... and one should note here that reductionist viewpoints emanate equally from the social and human sciences as from the natural sciences, and are equally fallacious in those cases too. (Ellis, 2004, p. 7).

Physicalism is a reductionist principle which was introduced as a philosophical concept in the early 20th Century by Austrian philosopher Otto Neurath (1882-1945). The principle asserts that all that is real is only "the physical, the rational, the programmable, or whatever is given primacy" (Jones, 2000, p. 14). In other words, reality does not exist beyond the physical realm or "beyond what can be explained by known physical laws" (Gebelein, 2013, p. 159). When we approach phenomena from just a material or external perspective with no thought, awareness or understanding of its internal spiritual reality, we cannot know its infinite properties. If we subscribe to this ideology, we risk viewing human existence as being potentially spiritually barren.

The Work provides a counter to this, which is why its esoteric knowledge has such practical application. While both the Fourth Way and Beyond the Fourth Way teach that we are as Ellis states, much more than atoms and molecules, they do something else as well. They demonstrate that the material universe and the non-material spiritual universe, while co-existing with one another, have different realities because they operate under different laws.

> The ancients clearly understood that an infinitely wise, caring and loving living essence is the origin of all things. From this Divine origin came forth a series of worlds of increasing density, and the entire creative

> process terminated in the physical universe. What we see around us are the end-effects of a vast series of interconnected processes reaching from world to world, and finally coming into visible and tangible manifestation. Behind even the most commonplace objects there are hidden layers of meanings, values, energy processes and states of consciousness. The awakened and refined human mind can undertake an inward journey into higher worlds, and there discover the reason for things, the cosmic functions of things and the vast meanings they represent. Most orthodox scientists completely avoid this kind of investigative research and provide many theoretical reasons why they do not pursue it. They prefer to manipulate the end-effects of the creative processes and remove themselves from the study of causes. (Groves, 2000, p. 9).

My experiences have always occurred unexpectedly and without the use of psychedelics or other behaviours such as for example fasting, meditation, chanting, reciting mantras or religious scripts and texts, drumming or dancing, which can often induce an out-of-body state. Irrespective, when such things occur, they enable us to experience the inner reality of our being and the spiritual reality of the universe in which we live. Because of their educational nature, we understand through lived experience that not only are we spiritual beings living in a physical body and a multidimensional universe, we learn that the world around us contains "hidden layers of meanings, values, energy processes and states" (Groves, 2000, p. 9).

Many people spend hours discussing with one another the various intricacies of one religion or another, of one system or another. They theorise about the latest best-selling book or emerging author promising new methods and techniques guaranteeing illumination and insight. They weigh up the merits of a new psychological method they might have encountered, and they devote time to intellectualising and theorising about God. They try to determine what God is or is not. They wonder if God is gendered, black or white. They postulate on whether God and/ or a belief in God is simply a salve for the masses or a means of social control and manipulation.

A telephone conversation between myself and an acquaintance at precisely the time that I first wrote this chapter was synchronistic. As the conversation was coming to an end due to my having to attend an appointment, the woman said to me before ending the call, "When you come back ring me and you can tell me what you think about God." In exasperation I instantly replied, "You can't think God, you have to feel God."

The finite mind cannot understand the Infinite and though some may disagree, no one can establish an emotionally loving and affectionate relationship with Higher Life by intellectualising, theorising or rationalising about it. One of the reasons why this is so, is because ultimately these are *thoughts* rather than a *state of being*, which is something very different.

> The observation is this. All people inherit and acquire every substance and energy that is necessary for the transformation of their being into a spiritualised formation that can become a useful and necessary aspect of timeless, eternal realms, but incredibly few take the trouble to do anything about it. Far too many assume that simply because they are human beings, they have a soul and when they die this 'soul' goes tootling off into blissful realms of paradise. Others assume that because they are religious or subscribe to some set of beliefs that their eternal future is secured. These are dangerous dreams that lull sleeping people into

> a false sense of cosmic security, and they make no serious efforts to change themselves from 'seeds' into fully developed beings. (Groves, 1996a, p. 3).

Participation in the Work and work-on-oneself requires more than rationalising its esotericism, reading lecture notes and books, and discoursing endlessly with others about the finer details. It requires living the teaching and incorporating it into our lives so that we *experience* life differently. Crucially, it involves establishing an emotionally loving and affectionate relationship with God and Higher Life. "Love and not intellect is the fundamental condition for the operations of celestial existence. Until a person learns to love the highest realities there can be no heavenward growth beyond the spiritual state" (Groves, 1997a, pp. 7-8).

In writing this book it was never my intention to produce an intellectual regurgitation of all the technical aspects of the Work; there is a plethora of books already written by Fourth Way teachers and writers which communicate these (i.e., Dr Maurice Nicoll's *Psychological Commentaries on the Teaching of Gurdjieff & Ouspenski*, Samuel Weiser, Inc., 1996, and any of Dr Philip W. Groves' publications, Triam Press). Nor was it Dr Grove's intention that the book be written for students who attended his School; indeed it was never conceived as such. I always knew that the book's audience lay outside the confines of the classroom as it did those who were once its students, and as its author my aim is to combine esoteric knowledge and lived experience to demonstrate in a practical way, how we can effect self-transformation and growth in our day-to-day lives with the assistance of the Work and Higher Life.

In concluding this introduction, it is my hope that the reader will be encouraged and challenged to not only keep reading, but to perhaps reflect on their own existence through the lens of an alternate transpersonal or spiritual perspective if they don't already do so.

By way of format and content, each chapter of the book is dedicated to a particular theme. For example, Chapter 1, *Toward Eternity*, which follows this introduction discusses my childhood from the context of how particular spiritual experiences were pivotal in providing the foundation of my worldview and spiritual belief-system. Chapter 2, *School Remembrances,* discusses my introduction to the School I attended and to its teacher, Dr Philip W. Groves, while Chapter 3, *Know Thyself,* discusses the context of the Work.

The fourth chapter, *Technical Terms,* provides an explanation for some of the terms utilised throughout this book, and Chapter 5, *Work Basics,* introduces the three foundational principles of the Work: Self-Observation, Non-Identification and Self-Remembering. Chapter 6, *Faith,* introduces one of the three functions of the essence, Faith, and discusses its relationship with spiritual growth and development while Chapter 7, *What am I?* discusses the importance of understanding what we are as opposed to who we are. Chapter 8, *Intentional Effort,* discusses the necessity of effort, the relationship of that effort to awakening from the sleep-state, and love for the Work, while Chapter 9, *A Word About Death,* explores the relationship between death and the afterlife and the possible purpose for our being born which loops back to the Work. Chapter 10, *A New Way,* discusses the emergence of a new way of being in the world.

Chapters 9 and 10 are additions to the original manuscript which I feel necessary to include because of new understandings gained after the manuscript was initially written. Following this is another addition, an Afterword entitled *A Final Note*. It is not an element of the book proper and is an account of the closing of the School. I feel it important to include because it signified the moment in time when my umbilical cord to the School and its teacher was cut, and I strove out into the world, alone, yet never lonely. Dr Groves had performed his duty with great love, dedication and humility to all the students he had known, and now it was time for this one, me, to do the same.

For those new to the Work or encountering it for the first time, be ever aware that "one of the greatest hindrances to a full view of reality is the universal human malady of psychospiritual sleep" (Groves, 2000, p. 9). We can never forget that the sleep-state is the barrier which prevents us all from seeing the inner reality of ourselves and of the spiritual pattern of life. To this day, I thank God that not only was I made aware of this from a young age, but that the lived experience of that has enabled me to truly understand the perilous nature and reality of the human condition.

1

TOWARD ETERNITY

> There is an indefinable mysterious power that pervades everything. I feel it though I don't see it. It is this unseen power which makes itself felt and yet defies all proof, because it is so unlike all that I perceive through my senses. It transcends the senses.
>
> M. S. Deshpande, (Ed.), *The Way to God: Selected Writings from Mahatma Gandhi*, pp. 6-7, 2009.

In this chapter I discuss my childhood from the perspective of how other-worldly spiritual experiences shaped my worldview and spiritual and religious belief-system, and how they laid the foundation for how I have lived in the world ever since. The disclosure throughout of personal experiences from my private life, in addition to revealing aspects of the multidimensional universe, serve also to demonstrate multidimensional reality. However, they illustrate something else as well; that there is an undeniable intelligence permeating the universe which seems to purposefully orchestrate such events.

While some may see my previous statement as poetic license, my recollection is that whenever I experience these events, I always feel that they're being done *to* me by something external to myself, and that I'm simply a participant in what's unfolding. The events are random and spontaneous, and I've never had any control over when and how they occur. Because of them, I have been educated about the spiritual pattern of reality and its multidimensional framework.

There are several threads which coalesce into the tapestry of my being in the world and my engagement with the Work. That tapestry accords with Charlene Spretnak's participatory universe and the expansion of that term in her book, *States of Grace* (Harper-SanFrancisco, 1993). In this book, Spretnak, American author and ecofeminist, discusses how and why the impact of modernity has resulted in a lack of meaning and interrelated problems for contemporary Western societies. The solution? The direct perception of a larger reality, and an understanding of the interconnection between all things spiritual and material. Referring to what we are in our minute parts, "existing within a vast network of creative dynamics" (p. 22), Spretnak urges engagement with practices which will awaken the mind to the reality of such participation. What practices do we engage in, and how will the mind awaken? For me, that constitutes work-on-oneself.

Although I was born in Australia, from three to eleven years of age I lived with my immediate family in the small town of Boroko, a suburb Port Moresby, in what was then known as Papua New Guinea. My relationship with the country that hosted my birth is evidenced by the one and only memory that I have of that time.

I was standing on the lawn in front of our house in Sadleir, Liverpool, in the Western Suburbs of suburban New South Wales. I recall that the house was white, glaringly so. And hot, it used to be so hot. A small cement porch graced the front door, steps led down to a gravel drive. My mother, who looked impossibly tall to me, was talking to our next-door neighbour while I played with the neighbour's daughter. I was bare-footed and wanted to walk across the lawn, a task unachieved due to the prolific number of prickly bindies growing in the grass. Then I recall the daughter carrying me, my arms around her waist and lots of laughter, then the sharp prick of the bindies (Soliva Sessilis) under my feet as I slowly slipped from her arms onto the grass. And then light, impossibly, blindingly, brilliant light .

This is all I remember of my very early childhood in Australia. I was two and a half years of age and it is still my only memory and one tangible link I have with the country in which I was born. After that, and for the formative years of my childhood, my worldview and spiritual belief-system was shaped by the lived experience of the reality in which I existed, a reality framed by a country and culture foreign to that of my birthplace together with a host of diverse other-worldly events. These included out-of-body experiences, dreams that were more than a dream, and direct interaction with disembodied or spiritual beings. I never questioned any of the events that took place because it was just how things were. At the time I couldn't categorise or name these things as such, that came later when as an adult I was taught what they were.

Reflecting on the totality of my childhood, I can only conclude that in addition to being socialised into multidimensional reality, I was born into the ordinariness of it. Whether we acknowledge it or not, or are aware of it or not, we are all born into multidimensional reality. We are all spiritual beings living in a material body, and what my childhood taught me was the simplicity and truthfulness of that basic fact.

As a child I felt that there was something special about life, something tangible yet intangible, which was an annoyingly constant contradiction. I felt that life was beckoning me to itself, was whispering to me in quiet moments, was touching me in a certain way because it wanted me to know it. Seemingly inert objects were vibrantly alive. Trees were not just intricate and sometimes perilous structures on which to climb, they were living beings. And to the mind of a small child, the magnificent sunsets which lit the darkening sky when dusk fell were magical events which filled me time and again with a sense of absolute wonder.

Never was I more aware of this than when I found myself roaming the jungle with my father (who was fond of hunting) or swimming in the shallows of fast flowing rivers, with crocodiles silently poised in anticipation watching from opposite banks. The sunsets, the stars at night, the moon, the garden of our family home, all of it was alive because it was impregnated with something, an energy, which made it so. Trees, seemingly sedentary structures oscillated with an energy which I could feel passing through my body. The jungle, a fecundate profusion of growth, was to me a living, breathing entity. Walking among towering trees and sitting between their massive roots filled me with an overwhelming sense of awe and mystery at what had created them. Looking into the starry night sky filled me with a yearning to see beyond those stars, to know their very origin, to stand in the place where they had been born. At all times I felt that *life* was pressing itself against me to such an extent that I felt I could breathe it, taste it, take it into myself and feed from it. Life was profoundly meaningful in a way I couldn't understand yet could feel deeply, inescapably within myself.

I would often wander by myself through our rambling garden at dusk, absorbed by the colour of flowers, shrubs, and even the grass that massaged my bare feet. Brilliantly coloured butterflies would float past, seemingly drunk on an abundance of nectar, while beetles and other insects, iridescent with colour that took on a different sheen in the soft twilight, dotted leaves of various plants and bushes. And at a twilight, that moment of the day when evening approaches, the whole garden would be bathed in the most beautiful hue and tones of the setting sun; golden-orange, pink and red light would suddenly flood the garden suffusing it with an unearthly glow. There were times when I fell almost into a swoon, and at those moments I was not only deeply moved by the beauty of everything around me, but also felt more than ever that I lived in an unearthly paradise.

And, amongst all of this 'life' there was another constant. It was the sense that something unseen, something vast and profound was continually pressing itself against my awareness wishing and desiring to be known. This was so strong that it seeped into my body and my mind, leaving an indelible impression which created a constant internal tension because by its very nature it urged me to seek it out, to unite with it, to become one with it. Writing of that period now I see reflections of my childhood in the words of English novelist and Anglican clergyman Charles Kingsley (1819-1875):

> When I walk the fields I am oppressed now and then with an innate feeling, that everything I see has a meaning, if I could but understand it. And this feeling of being surrounded with truths which I cannot grasp amounts to indescribable awe sometimes ... Have you not felt that your real soul was imperceptible to your mental vision, except in a few hallowed moments? (Kingsley, 2011, pp. 59-60).

Kingsley used the words 'hallowed moments'. Rather than hallowed moments, it seemed to me that my childhood in its entirety was hallowed.

In addition to frangipani, hibiscus and banana trees, our extensive garden was graced by two towering, gnarled and aged mango trees. Prolific producers of the fruit which often fell to the ground, their branches when scaled provided an opportunity to view the world from lofty heights. I remember the sensation of the bark against my skin, the acrid smell of stink bugs as I brushed up against them and feeling 'butterflies' in my stomach as the branches below me moved and swayed in rhythm with the wind.

I loved those two trees and often fancied they were 'protectors', watching over me and keeping me safe from harm. Some may perceive this as the simple fancy of a child's mind, but whenever I was near them I felt a kind of internal hum emanating from them, and I often felt they were communicating something to me but I could never understand what that was. Interestingly, in all the times that I found myself in their branches, and, when much to my mother's horror I would walk on the roof of our family home, I never once fell nor ever feared I would.

I particularly remember the following occasion which occurred when I found myself standing in front of one of them. I looked at the bark, noticing the colour and texture, and watched ants crawling along unseen pathways. How was it they never bumped into one another or other objects in their path? I could feel the heat of the sun as it shone down on me and could smell the frangipanis fragrance heavy in the air. Suddenly I became aware of the sensation of my heart beating forcefully in my chest together with the sound of my blood rushing in my ears, and then the world folded in on itself, taking me with it.

I tilted my head backward, my eyes moving up the towering trunk to its great green canopy and suddenly I felt that the tree wasn't a tree. On the outside of course it was, I understood that, but it was something else as well; what it really was, was not its appearance. I stepped toward it and opening my little arms wrapped them as far as I could around the trunk, which was so wide that I felt as though I was simply standing with my arms outstretched.

And then something remarkable happened. As I held the tree, I felt that it was holding me and though to this day I still cannot explain it, I felt that something within me *loved* the tree. I felt its vibrational frequency blend with mine, and it seemed that the physical world around me dissolved. It was like two notes being played on a musical instrument but with only one sound being heard. The material world fell away and I felt myself melting into the tree, becoming one with it, and then I was the tree and the tree was me and there was no separation between the two.

But there was something else. I could feel that what made the tree *was* the tree, and that what that source was, was also encased within it. Not only was the tree a physical manifestation of that source, it also served as an appearance for an interior meaning; it was a coded message. As the essence of the tree blended with mine, light filled my mind and then I was overcome by the feeling that in addition to being loved by something vast, which resided outside and within me simultaneously, I was deeply cherished as well. And I loved that which loved me, innately, completely, truthfully.

All this I knew in an instant, yet I don't know how I knew it, I just did, and it was like so many of the spiritual or other-worldly events that have occurred throughout my life since that time. I lived in two worlds, always seeing the external representation of things while simultaneously feeling or being shown their inner reality. I remember this event with the mango tree vividly, as I do all those that have occurred throughout my life, and I can recall every detail. The reason I can remember them so clearly is not only because the event stood outside the realm of the usual day-to-day events and experiences which occurred in the world of a child growing up; it was the combination of the event and my feelings in response to it, because it was these which confirmed the reality of what had occurred.

There was another factor which influenced and determined how I was predestined to live in the world; an ability to simultaneously see and feel material and non-material phenomena as varying vibrational frequencies. This taught me that every object, every 'thing' oscillated at its own vibrational frequency, and that for some obscure reason there was something within my being which enabled me to know and experience the world in this way. It was very difficult because I didn't like seeing things around me continually 'shudder', so as young as I was, I trained myself to focus on *feeling* their frequency instead. To do this, I would focus on the centre of whatever object I was looking at, concentrating with all my might, which was very hard for a small child, until what eventually remained in my vision was a one-dimensional image, flat, solid, and unmoving. When this external seeing switched off, an internal seeing switched on, together with an ability to *feel* the vibrational frequencies internally, instead of seeing them with my physical eyes as I used to.

This enabled me to metaphorically look out (to see the material world with my physical eyes) and simultaneously to look in; an ability which enabled me to feel the internals, for want of a better term, of whatever was around me including the environment, people, animals and what is defined as non-material, other-worldly or spiritual phenomena. This translated to my being able to 'see' objects in my mind which was not dissimilar to looking at a picture.

I can remember even then that I knew that life was revealing aspects of itself to me not unlike the turning of a page in a book reveals its unfolding story to the reader. Although I didn't understand *what* was at work, I knew irrevocably that something *was* at work. My lived experience of these phenomena was the evidence for it, and because of their impact I was born into the experiential understanding of the existence of another reality, an internal spiritual reality, which existed beyond that which I could see, hear, feel, smell and touch.

As my childhood progressed, I came to understand that these

two distinct realities, these 'places', not only co-existed with one another they interpenetrated one another. While both realities had distinct phenomena which defined and delineated them, the boundary between each was permeable; hence, 'reality' was multidimensional.

One type of phenomena was detectable by the physical senses and easily discernible. This was the reality of the material universe in which I lived as an embodied being, in which I lived as 'me', a living, breathing, human being. The other type of phenomena was made visible by a different kind of sensing completely unrelated to the physical senses. This was the reality of the non-material spiritual universe, although at the time I didn't know this 'place' even had a name. All I knew was that it existed and that it seemed to go on for ever and ever, it never stopped because it never started, it just *was*. This vastness stood behind all that I could see with my physical eyes and experience with my physical senses.

The only way that I can describe this is that I saw this spiritual reality with eyes inside my eyes and experienced it with internal non-material senses quite apart from my five physical senses. It was as though the reality of this non-material universe was a canvass and the reality of the material universe was a scene painted on that canvas. While this duality of seeing granted me simultaneous entry into the reality of both universes, it also bore with it a responsibility that constantly pressed upon my awareness; the responsibility to know and acknowledge both yet to not speak of either. So, for many years I never did.

What prompted this decision was something I felt intuitively. I couldn't understand or comprehend how I knew this, because it wasn't my mind that was thinking it, it was another part of me that was feeling it. I just *knew* I could never speak of these things to anyone and so from a very young age I learnt to keep my own counsel. Because I followed my intuition, I not only kept what may have been unwanted attention diverted away from myself, I learnt to metaphorically *feel* my way through life rather than *think* my way through life.

This intentional withholding took on greater significance as the years passed, because I learnt that appearances were deceptive; not only was the visible world other than what it appeared to be, but people were too. I have a peculiar ability which enables me to 'read' a person, and I frequently experienced occasions whereby what a person said to me contradicted what their inward thoughts and intentions were, which made me cautious.

I also dreamt a lot as a child and was frequently aware when I did so that I was simultaneously both observer and participant in my dreams at one and the same time. I used to feel I was in two places at the same time; I was in the dream, and I was watching myself in the dream. I felt the dreams I had were important, though I didn't understand why, and that they were trying to tell me something, but I didn't know what. The Work has taught me that dreams are comprised of a symbolic language which as a child I couldn't possibly understand. Even so, I *felt* that they were communicating something to me and it was that impression which always remained after I had woken.

I was also aware of something else. I would often wake during the night to find who I named 'the shinning people' standing around my bed. I gave them this name because they shimmered and shone as if illuminated by an inner light. Their garments, flowing robes of various colours, looked like they were woven from hundreds of tiny stars and were reminiscent of sunlight playing over water. I was unafraid of them and though they never moved toward me directly I always knew their intent. In addition to my waking and seeing them standing in my room, they would come to me in my dreams and here I would often find myself walking with them in unknown lands of great beauty, with beautiful buildings, and mingling with others like them.

They didn't use their mouths to speak as it was their thoughts I could hear in my mind, and always on waking, and in addition to feeling as though I had been 'away somewhere', it wasn't their thoughts I recalled but the energy that seemed to comprise those thoughts. Because of my difference to those around me, and because I knew I could not speak of my unusual experiences, as a child I often felt quite lonely and isolated. But those feelings were always softened by my interactions with the shinning people. They emanated warmth and love toward me, and I knew my wellbeing was their utmost concern, which gave me great comfort. I also knew something else intuitively, that they understood me as they did how I lived in the world.

When I was in my twenties and not long after I had entered the Work, I had a dream which further demonstrated my inner relationship with the shinning people. Somewhat overcome, as I often was by my dreams, I had written it down and given it to Dr Groves to read. When I next saw him, he told me the meaning of the dream (which follows in italics):

> I found myself on a distant planet which was hot and arid. There was great commotion and people were running hither and thither because a new world was being born. The ground started to heave and shake violently, and a great roaring wind deafened the ears and whipped up dust making it difficult to see. Through the haze created by the dust I could see people running to nearby waiting spaceships to escape the mayhem.

Suddenly above the noise, I heard my name being called and looking into the distance saw a group of people standing under the branch of a huge tree. I began to run toward them, but losing my balance fell frequently due to the violent shaking of the ground. The intensity of the wind was so great that at one point I lost my way due to being temporarily blinded by the dust which had now been whipped up into a choking cloud. At precisely this moment, they called to me again, "Michele, Michele" and by orientating myself to the direction of their voices, which were clear and distinct above the howling of the wind, I was able to find my way again.

The heat was intensifying as was the wind and the shuddering, dust blew everywhere; it was absolute chaos. To my right I was suddenly aware of a great illumination. Turning toward it, I saw a huge and brilliant sun rising over the horizon. I watched in amazement because it was so close that I could almost touch it. As I looked into it I could see that it consisted of circular layers or spheres, one within the other. The most central or inward layer was white, yet a different white to the colour perceived by physical sight. Around this central white area was another layer, which was lemon-yellow in colour. The colours are difficult to describe because they were *shades* of colour and there aren't words available to suitably describe or capture the subtle variations which were present.

I thought to myself that if I simply reached out to the sun I could touch it. At that moment the voices called to me again, "Michele, Michele" and hearing them I turned away from the sun and once more continued toward them. I had been walking for some time now and my legs were growing heavy as I struggled to navigate the shaking and shuddering earth. They were still far away and I was tired but intuitively I knew that they knew this, and at that moment called anew, "Michele, Michele!"

As I continued toward their voices the light from the sun was blinding but knowing I couldn't stop, compelled myself forward. Although I was tiring, they did not stop urging me onward either. The ground heaved and shuddered, the wind, whipped into a frenzied howling roared around me, the heat was intense and the light from the sun blinding.

On and on I went, pushing myself, struggling against myself, knowing that I had to get to them. And then suddenly I was there in the shade under the branch, away from the light and the heat and the roaring wind and its dust.

I fell exhausted to the ground and looking up at them, was only able to see their faces and forms in a muted way. It was as though I was looking at them through opaque glass, yet despite the separation I knew that they knew me and that they had helped and guided me through a difficult and precarious psychospiritual event. I knew that these people were my spiritual community, that they were the shinning people who used to come to me when I was a child. And with that knowing, I awoke. *(Author's journal notes, 14 April 1996).*

The shuddering of the earth is vastation. The great wind is the same as that which appears in the spiritual world which blows away evils. The individuals leaving by spaceship are the evils (negative beings) which are in fact being blown away. They think they can escape of their own volition however they are mistaken. The Divine is the one and only the one who can remove evils and negativities from us. The great sun is the Holy Sun Absolute. You see a core of blinding white light surrounded by another light. The inner core is referred to by the ancients as Yahweh, the outer circle of light is the personified human. The spiritual sun consists of many levels and you perceived those levels. The people calling you are your spiritual kith and kin. This is the birth of a new Heaven and a new earth and is akin to something from Revelations. Something in you is dying, something is being born. This is the foundation of something great.

(Dr Philip W. Groves, personal communication, 22 April 1996).

There is another aspect to this issue of one's spiritual kith and kin that bears mention. As a child and as young as I was, I intuitively felt that the shinning people knew me before I knew them. There was something familiar about them and I can still remember feeling that I belonged with them, that I was a part of them and that I too came from the same place, that 'other place', that they did. I also felt intuitively that they were my real family rather than the people with whom I shared my embodied life. For reasons beyond my comprehension, I knew they recognised me and that they had taken it upon themselves to make themselves known to me. The beauty of childhood is that one experiences life without the blinkers of an intellect, a thickly layered personality and an overly dominant, rigid ego, or 'I' which can often interfere with what one intuitively feels. I knew in my childhood heart the truthfulness of what was taking place; I knew it with a purity that was as great as it was humble in its simplicity.

In my childhood I had felt that this other place, like the material universe I lived in as an embodied being, was characterised by an immeasurable vastness and that it had no beginning and no end. I also knew, again intuitively, something else. Time as we knew it, linear time, didn't exist in this place; 'time' didn't exist at all. Many years later, after undergoing a near-death event as an adult which literally catapulted me out of my physical body and into the spiritual universe (which I had since come to know and understand as 'that other place' I had been so aware of as a child), I experienced this timelessness again. But by then I knew what the timelessness was; eternity.

When I was a young schoolgirl, my early education was gained by attending the local Catholic primary school where religious studies were mandatory, as was attending church on a regular basis. Having been baptised into the Roman Catholic faith, I remember attending church (for Sunday services and feast days), a foreign language spoken by the robed priest (which I learnt later was Latin), aromatic clouds of sweet-smelling incense (which instilled a lifelong love for it), making my first Holy Communion and visiting the confessional (a requirement which I never quite understood).

I recall in particular an event which took place in class when our teacher, a nun, was explaining that when we die we go to heaven. Because I thought a lot about death and the afterlife as a child, and because I wondered what we did when we got there, I asked her the following question, "What do we do in heaven?" She looked at me with a startled expression on her face. I could feel her bewilderment and fear as she stood there trying to answer my question. And then I saw that she couldn't answer the question because she didn't know what the answer was.

As I sat in my chair waiting, she became more uncomfortable. She couldn't tell me, yet wasn't she was supposed to know? In my mind, as young as I was, I reasoned that she was a nun; she was meant to understand about God and heaven and all those things. Heaven wasn't just a place that people went to when they died, it was a place where other things happened too, and although I didn't know what those things were, I felt the truth of it intuitively; I just knew.

I felt that I lived in a world that had come from a special place and if that place wasn't heaven, I didn't know what was. The shinning people came from that place, it was their home, and on some level I knew that 'place' was my home too. Was that heaven? Is that what it was called? These were the thoughts in my mind when I asked the question, because I knew, again intuitively, that when we died we *did* something, but I didn't know *what*; we didn't just disappear. I needed to make sense of my other-worldly experiences and having my question answered would have helped me do that. But the nun didn't know, and I reasoned that if a nun couldn't help me, who could? Fifteen years later, my question was answered and I understood then why the nun didn't know; she hadn't been taught.

Throughout my childhood I remember being confused by these thoughts and feelings and wondered why it was only me who was affected in this way, why it was only me who felt and knew there was more to my existence than just being born and living. The people around me seemed content with their existence but I wasn't. Why did they live one way and I another? Why didn't we all live the same? They went on living day in and day out, doing this or that. In observing them for example, I knew what my mother's reactions to various happenings and events would be as I did those of my father and brothers.

Everything was the same, they were the same, the people around me were the same yet life was calling to them too, was beckoning to them as it did to me, was touching them as it touched me. Why couldn't they see, why didn't they know? My mind was alive and alert, and even then I was questioning the status quo, was challenging what people told me to be the truth, was challenging myself. As the years passed, people came and went in my life, one event after another occurred, and all the while my dissatisfaction, my discontent with existence as I knew it gained momentum and intensified.

Death was something I frequently thought about as a child and I recall that the concept of being dead always seemed confusing. This was because my experience of death contradicted what my parents, the nuns from the school I attended and the people whom I had contact with said about it. The reasoning that would take place in the mind of a small child was thus. If people died and went to heaven, and if you could only go to heaven if you were dead, and if you wanted to see or talk to a person who was dead because you missed them, then you would have to die and go to heaven too because that was the only way you could do that.

But if people were meant to be dead, and if being dead meant that they disappeared and couldn't be seen any more, how come they could still be seen? And what about the shinning people? I knew they weren't 'alive' because they didn't live like everyone else around me did and they didn't live in the world that I lived in; they came from somewhere else. They were different, but because they were different didn't make them any less real. Did this mean they were not dead after all? What then did being dead mean?

This occupied my mind a lot and I can still remember reasoning that it was like the car I used to travel in with my mother. We got in, we drove somewhere, and then we got out. My mother's actions animated the car making it move. Without her engagement it didn't move. If the car was like a person and my mother like the life force, then without that life force the car wouldn't move, wouldn't be animated. The car was still there but it was inert, empty. My mother was still there, she just wasn't in the car. That made sense to the questioning mind of a curious child.

When I was barely into my teenage years, and we had returned from Papua New Guinea and were now living back in Australia, my mother told me about my great-grandmother and her extraordinary mediumship ability known in spiritualism as the 'gift'. I had grown up listening to ghost stories and all manner of odd happenings and strange coincidences, however it wasn't until I learnt that great-grandmother Rees was a spiritualist and a trance medium that for the first time, I was able to conceptualise what might have been the cause for my experiencing life as I did (but which I found out years later had nothing whatsoever to do with it).

My mother introduced to me terms I had never heard before, such as 'spiritualism' and 'spiritualist', the 'other side', a 'sitting', a 'séance', 'trance mediumship', 'ectoplasm' and 'rapping'. She also entrusted me with my great-grandmother's crystal ball which to this day remains in my keeping. Despite my mother's disclosure about my great-grandmother, I never told her about my experiences, that came later when I was an adult, and then only very selectively. However, I think on some level she knew or sensed that I was different; she used to often say that I reminded her of nanna Rees, despite never clarifying in what way.

In addition to being the Reverend for her own spiritualist church in Glebe, an inner-city Sydney suburb, great-grandmother Rees had been a Deaconess in the High Anglican Church in England, her country of birth. As far as I know very little evidence of her life remains, other than a few faded photographs, some handwritten notes, and an oral record of her clairvoyance and mediumship entrusted to me by my mother. My mother told me that prior to her death which had followed a long illness, she had been 'told by the spirits' which herbal remedies would heal her ailing body. According to my mother her response was, "I want to go home." Some weeks later she died and was buried with little ceremony in a small, fenced plot in a local Sydney cemetery on a sloping hillside facing the ocean.

One of the stories of great-grandmother Rees's mediumship particularly fascinated me and at the time of its occurrence had both astonished and frightened my grandmother. As a young child she had peeked into a curtained alcove where her mother received callers in the family home who came for a sitting. To her shock her mother was speaking in a foreign language to the woman who sat opposite, and what was more, the woman was responding in the same language. English born and bred; her mother knew no language other than her mother tongue.

My grandmother had told me that throughout the years she had heard her mother converse in many different languages, all fluently. She thought it funny when a man's voice came through, at how the timbre would change. Her facial features would change as well, seemingly taking on the characteristics of the disembodied being who was speaking through her. When the interaction passed, she had no memory or recollection of what had transpired. My mother told me that my grandmother had inherited the gift, but that it was dormant inside her because of her fear of it. It had missed my grandmother's sister, as it had my mother, coming instead, according to my mother, to reside in me.

My understanding of this inherited gift from my mother's maternal linage has changed since I first became aware of it. Life experience, contemplation, solitude, self-examination and my engagement with esoteric work has considerably altered the initial context in which I had positioned the other-worldly phenomena associated with the gift. Once I would have defined the dead as 'those on the other side' without really knowing what constituted 'the dead' or the 'other side'. What did these terms mean? What did *the dead* mean? What did *the other side* mean? I was dissatisfied not so much because the terms had lost their initial currency, but because I knew my understanding of them was inadequate.

Contrary to the explanations offered by spiritualism for the dead and the other side, I knew that a vastness extended beyond the reality defined by my senses, and I knew that that same vastness stood above and beyond the realm of spirits particularly associated with spiritualism. The foundation for my awareness of multidimensional reality had been laid in my childhood, but in growing to adulthood I has become aware of a disconnect between the events that occurred *then* and my understanding of what their purpose was *now*. I felt deeply that something intangible yet also tangible stood behind all that I could see and that in an almost perverse way my senses contrived to work against me to produce a screen which obscured my seeing what that was.

Paradoxically, I also had eyes within my eyes, an ability that enabled me to see beyond that screen and while this duality of seeing has remained with me my entire life, at that time I was utterly consumed, driven almost to a kind of inner madness to know the source from which all this stemmed. If I knew that, then I could make sense of what was happening to me.

And so in silence, first as a child and then as a teenager, I endured a merciless hunger and an unrelenting yearning, and it was these things which characterised the inner reality of my life at that time. It is difficult to convey in words the intensity of the inner tension I continually felt during this period, and it was not until I encountered and entered the Work at the age of twenty that this inner tension was ameliorated (see Chapter 2, *School Remembrances*).

The experiences recounted in this chapter grant a glimpse into the multidimensional universe we are all born into. In his Ode, *Intimations of Immortality from Recollections of Early Childhood* English Romantic poet William Wordsworth (1770-1850) laments the loss of that time when as a child "every common sight ... did seem apparelled in celestial light" and although "Heaven lies about us in our infancy, shades of the prison-house begin to close." Then, as the youth travels "daily farther from the east", at length "Man perceives it die away, and fade into the light of common day" (1884, p. 24). Wordsworth's lament of loss contrasts with Kingsley's hallowed moments as it does with my own experience of awe and wonder at the ineffable, which is as gently and consistently present now as it was when I was first born into it all those long years ago.

2
SCHOOL REMEMBRANCES

> The Western mind must be willing to open itself to a reality the nature of which could shatter its most established beliefs about itself and about the world. This is where the real act of heroism is going to be. A threshold must now be crossed, a threshold demanding a courageous act of faith, of imagination, of trust in a larger and more complex reality; a threshold, moreover, demanding an act of unflinching self-discernment.
>
> Richard Tarnas, *The passion of the western mind*, p. 444, 1991.

As I grew from child to teenager, the persistent feeling that there was something precious just within my grasp yet always out of reach intensified such that it was a constant presence within me. I felt an inner restlessness and an intense dissatisfaction with my life as it was, and I was often frustrated and melancholic as a result. My parents saw me as a precocious, tempestuous, and rebellious teenager and I don't hold them to account. I was like that, but it was because I was unable to articulate my feelings or be understood. I *knew* there was more to life, but I didn't fully understand or know what that was. And I *knew* there was another way to live but I didn't know how to live that other way. And what added significantly to my frustration, was the fact there was no one who could show me *how* to live that other way.

When attending secondary school, I desperately wanted to

learn, but the difficulty was that what I was being taught wasn't what I wanted to learn, apart from the writings of the great English playwright William Shakespeare that is. My recollections from that time are not entirely pleasant; however, one light in the proverbial darkness was my receipt of an old cloth copy of a volume of Shakespeare's plays. Edited by Howard Staunton and now dog-eared with brittle and yellowing pages, some of which are missing, it was gifted to me by a close friend. Originally the treasured possession of his father it has since accompanied me through my life, and though it sometimes maintained pride of place on my bookshelf, more often than not it has rested on my bedside table.

The conundrum of wanting to learn yet not being taught what I wanted to learn was tireless. If the knowledge I yearned for wasn't to be found in the contemporary Western education system, where was it? It was an extremely difficult period of my life and I often wondered what was wrong with me. Why wasn't I satisfied with the way things were, why was I so unsettled, why did I always want more? As my childhood had been characterised by a persistent sense of discontent, so too it seemed would be my teenage years.

Always practical (some would say task orientated), I began to search for knowledge and having been an avid reader from a young age put my love for the written word to good use. Throughout my schooling years I voraciously read all manner of books written by a range of contemplative authors whose focus was the relationship between spirituality and self-development. These included for example the writings of Lebanese American poet Khalil Gibran (1883-1931), American psychiatrist and author Morgan Scott Peck (1936-2005), and English author Lobsang Rampa, (born as Cyril Hoskin, 1910-1981), who if nothing else introduced me to Tibetan Buddhism. A friend gave me a book written by Silver Birch, an American Indian spirit guide who spoke through Maurice Barbanell

(1902-1981), founder and editor of the London *Psychic News*, and on one occasion during my many forays into yet another second-hand bookstore, I chanced across books authored by John Ruskin (1819-1900) English art critic and social thinker, and John Bunyan (1628-1688), English writer and preacher.

The odd thing about this period of my life was that although I felt that it was me who had begun to search, I always felt that it was something inside me which had precipitated the search; 'I' had nothing to do with it, 'I' was just following the lead. Though confusing, I knew I was inextricably bound to that search, connected to it in a way I couldn't understand. It arose from deep within me compelling me into action to seek out what I was looking for, even though I didn't know what that was. Adding to my perplexity was that whatever I read seemed to serve as a signpost which pointed me toward another direction which only increased the inner tension I constantly felt. Each book stretched toward something in the distance, intimating something, hinting, always, at something which lay beyond its pages.

Irrespective of the sensitivity of the writing, whatever I read didn't provide the knowledge I was looking for; but that was the issue, what *was* I looking for? What was this knowledge, and what was this thirst, this hunger that pulled me toward it? I didn't know what I was looking for, I only knew that when I found it, I would *know* that I had found it. Later when I entered the Work, I learnt the technical term which accurately described what was occurring, this 'that which you seek is seeking you' process; Magnetic Centre. Until that time, I had no idea or understanding that what I was experiencing was the very real psychospiritual dramatisation of something within me seeking conjunction with something outside and beyond itself.

After my secondary schooling was complete, my search continued and as it so happened I one day found myself reading one of Lobsang Rampa's books, *The Third Eye* (Secker & Warburg, 1956). I was twenty years of age and firmly convinced after reading it that if I wanted to pursue a spiritually contemplative life for myself, which I reasoned was how I might find the knowledge and way of life I was seeking, I would have to forgo my Western lifestyle and retreat into seclusion in a Tibetan monastery far away from the world.

At this time one of my brothers happened to mention in casual conversation that a lecture was being given by someone or other about the *Tibetan Book of the Dead* (Oxford University Press, 1960). I was instantly alert; surely this must be providence at work! I had never heard of this book but that didn't matter. It was about Tibet, the 'spiritual homeland' as I believed at the time, and that was enough for me. And so it came to pass that some weeks later I found myself accompanying my brother to an ordinary looking house in an ordinary Sydney suburb, to attend a lecture given by someone I didn't know about a topic which at the time was an interest of mine.

The front door to the house was open and light streamed forth from what I could see was a short narrow hallway, which had several closed doors on both sides which obviously opened into other rooms. My brother had manoeuvred himself behind me, and as I stood in the hallway feeling nervous and somewhat hesitant, I could hear voices and the sound of a television coming from one of the rooms. On my immediate left was a doorway and looking in I saw what appeared to be a loungeroom arranged with overstuffed couches, chairs and cushions, on one of which sat, rather smartly as I recall, an elderly man.

He must have sensed me standing there because he interrupted his conversation to briefly glance toward me, then he looked back at me again, this time more intensely with an expression on his face that startled me; it was a look of amazement. His expression took me by surprise, and when he realised that I was aware of his internal reaction he composed himself, drawing a mask down and over his face. He saw that I saw that too.

He immediately stood up, and I saw that he was tall and straight-backed. He wore a white collared shirt with a tie, dark trousers, and light brown shoes. His hair, which time proved was always a little sparse, was brushed neatly to one side; he reminded me of a kindly grandfather, but with a learned scholarly air. Then something rather odd happened. As I raised my right foot to step over the threshold to enter the room, I suddenly felt that I stood naked before him. Not naked in the sense that I was unclothed and physically nude, but naked in the sense that everything pertaining to me which was non-material or spiritually internal was revealed to him in its completeness.

I knew in that instant that he could see into me, that he could see processes at work within my being that I couldn't, and that he knew me in a way that no other person was capable of ever seeing, knowing, understanding or even comprehending. He knew more about me in that instant, in that one single look into my being than what I had garnered about myself in the twenty years to that moment in time that I had existed. And the reason he knew me in that way was because he had the ability to see the spiritual totality of the components of my being.

And then something else happened. As I stepped over the threshold and entered the room, I was overcome by a profound feeling that I had come home. I was *home*. It is impossible to put this into words because there are no words capable of describing what I felt. Although feeling somewhat self-conscious and awkward beneath his scrutiny, I found myself looking up into a kindly face with all-seeing eyes, which were studying me, assessing me, weighing what I was. I could see that though he was somewhat taken aback by the fact that I could see him seeing me, he didn't acknowledge that in any way. In fact, it was as though a silent or non-verbal form of communication was occurring between the two of us.

Suddenly I felt my hand gently clasped in the enclosing warmth of his, and in that instant I knew with every fibre of my being that my search for the knowledge I had been seeking was over. A stillness entered me. And a silence. And at long last that unrelenting inner restlessness that had driven me for so many years, that agonising unquenchable yearning, stopped, and the inner tension that had been my constant companion for so many years simply ceased to exist. And this is exactly how Magnetic Centre works; when it contacts the Work it dissolves because it's no longer needed.

I didn't understand a word of the lecture. The elderly man who introduced himself to me as Dr Groves, spoke about the Tibetan belief of what occurs at physical death, mentioning terms that I hadn't encountered before. Even though I had absolutely no comprehension of a word he spoke, I sat in that room utterly transfixed and quite literally fed on the energy encapsulated in the external form of the words that he spoke, which were a representation of something quite different altogether. I was aware that this was taking place, and though I didn't understand why, I did what I had always done when such events occurred in my life, I surrendered to them. I surrendered to them because my childhood had taught me to intuitively trust in what was taking place. My rational mind had enough sense at that time to realise that it didn't know very much at all about what life was about, and that if it was going to get on, it had to pay attention to what was taking place interiorly within my being, and externally to the other-worldly phenomena which occurred.

There was a kind of ethereal mist in the room, and I was aware that I was breathing in another way which was quite different to how I usually breathed. I felt like I was a great dry lake gradually filling with water, and it seemed to me that each word he spoke was code for something else. He spoke English yet I knew it was a different language altogether. I took notes in shorthand which when later transcribed contained many gaps, making whole sentences illegible. And it was no wonder.

Finding it difficult to concentrate during the lecture, and although I wrote furiously, I could barely keep track of what he was saying. I was not only experiencing a maelstrom of emotions, I felt that something inside me was at long last being fed and nourished by what I had been seeking for so long. I knew that I had finally found that 'other knowledge' that I had been looking for. I had no idea then that I had made contact with the Work or an esoteric Christian School which taught the principles and practices of work-on-oneself, because I didn't know that the knowledge I had found had a name, much less that it was capable of being taught; that understanding and recognition came later. All I knew at the time was that my soul had found its way home, and that it could feed, and now grow.

When the lecture ended and the small group of people filed out, my hand was once again clasped and gently held, and again I looked up into a smiling face with all-seeing eyes. Were they blue? "Please call and come and see me" he said, to which I somewhat bemusedly agreed (while wondering what on earth I could possibly talk with him about). And this was how I came to meet the man who was to have the most profound and lasting impact on my life as a teacher of the Work. From that moment I began to chart a course through life which despite my not knowing then, was to have far-reaching consequences for my psychospiritual growth and development, and for my eternal spiritual future.

Lectures were conducted on a weekly basis throughout the year and divided across two to three evenings. In attending a different lecture each week, a new world gradually opened its doors and ushered me inside. For me, the notion of existence changed from the secular to the religious at that time of my life. I realised that a person didn't have to enter a religious order if they wanted to have a relationship with God. I didn't have to travel away from home to a foreign land and I didn't have to renounce living a busy life to devote myself to the pursuit of a religious vocation. Furthermore, I came to understand that 'religion' was something quite different to what I had previously thought.

The etymology of the word provides insight into its meaning, which in turn assists in understanding how religion as an *activity* can be equated with spiritual growth. For growth to take place something has to be done, and in this context religion as an activity aligns itself with the Work and with work-on-oneself. The term is derived from the French of the same name, which in turn is derived from the Latin *religare*, meaning 'back, to bind': *Religare*, back, to bind (Latin) > Religion (French) (Skeat, 1993, p. 398, p. 584).

Religion then, is the act of binding back. What is it that we bind ourselves back to? We bind ourselves back to the creative source; we bind ourselves back to God. From understanding its root meaning I understood that religion in the context of the Work was an activity which not only had a specific purpose and goal, it had cosmic implications for my spiritual future. So, I began to bind myself back to God. I began to work on myself and I became an active participant in the renewing of my mind and the cleansing of my inner self.

I used to think that my journey began when I first encountered the Work but now, and after having lived it for over forty years, I know differently. In truth, my journey began when I was born into existence as much as it did when I took those first tentative steps in which I engaged with the other-worldly experiences in my childhood. It began when life blew its essence into my nostrils as it did when it touched my spirit. It began when as a child I looked up into the stars blanketing the night sky as it did when the afternoon sun dipped low and bathed the garden of my childhood in glorious colour. It began in the dreams I experienced then as it has continued to do in the dreams I've had throughout my life.

And the answer to what I didn't know then, which explained

the spiritual phenomena taking place, is one I know now; and it's mirrored in the same question that Carl Jung (1875-1961), Swiss psychiatrist and psychotherapist poses when he asks his reader in his autobiography *Memories, Dreams Reflections* (HarperCollins, 1995) whether or not they are related to something infinite. Jung explores this, observing that if we understand and feel that we already have a link with the infinite then we experience being in the world differently: how we see the world changes because our feelings toward it changes.

It seems to me that Jung has captured the essence of the meaning of life, which is connectedness to God through self-knowledge and conjunction with Higher Life, which plays a pivotal role in that relationship. Self-knowledge is the knowledge of what we really are in the clear light of reason and consciousness, rather than what we so often think and believe ourselves to be. The sleep-state renders us an illusion to ourselves and to the spiritual universe. Because of this illusion we don't know what we are, and we don't know what we are living in or for. Tragically, nor do we understand the cosmic purpose for our birth or what it is that we can become, if we want to.

Self-knowledge is what enables us to see what we are from an objective perspective. Seeing ourselves can not only be a difficult concept to come to terms with, but an activity which can create confusion, despair, even momentary hopelessness; when we see ourselves from such a perspective the disharmony of our inner being and lives is revealed. When this occurs, we don't know what to do with ourselves because we realise that we have no power with which to correct our faults or overcome our weaknesses, shortcomings and negativities. It is at this junction that the Work and work-on-oneself comes into play.

The Work is concerned with change of *being* which is quite different to the counselling and psychological work conducted on the personality. Esoteric work-on-oneself is concerned with spiritual rebirth, regeneration and transformation and it teaches us how to participate in the process of soul-growth and soul-evolution in a practical and useful manner. Self-knowledge and love for God are intimately linked and when both are combined, lay the foundation for a new way of living which is vastly different to that experienced by many people in contemporary societies.

Jung highlighted an indisputable fact when he posed his challenge, questioning the relationship of man as a totality existing within an Infinite framework. Is man related to the Infinite? Jung observes that if in this life we already have a link with the Infinite then desires and attitudes change. It seems to me that this is pivotal because it is much more than desires and attitudes that change; it is how we live our life that changes as it is our way of being in the world.

My attendance at that first lecture during which I was introduced to the Work was in many ways the first day of the rest of my life, and it heralded a new kind of study. But what was it that I was studying, what on earth did these terms which I was hearing for the first time mean? Terms such as Chief Feature, the Centres, the Higher Centres, the Law of Otherwise, the Scale of Octaves, Self-Observation, Self-Remembering, Non-Identification, Correspondences, Proprium, Ego, Essence, Consciousness, Personality, the Side Octave of the Sun, just to name a few, left me totally mystified. Despite my wide-eyed wonderment and complete lack of understanding of their meaning, and exactly what it was that I was meant to be doing, it didn't matter because I knew that one day I would understand.

I recall at the time being reminded of a similar experience I had when first starting primary school. I had been attending what was called then 'preps' which I think is the equivalent of kindergarten now. It was toward the end of the final term and I was due to enter first class the following year. Our teacher took us into the Year One classroom in order to familiarise us with the new room and layout of desks. There was a big green chalk board which stretched across one of the walls which had strange angular lines and circular marks on it, some of which had space between them while others were all clumped together. As the teacher pointed to each mark, the children were all repeating words.

I couldn't fathom what they were doing and then suddenly I realised that they were learning to read, and that the strange marks on the chalk board were 'words'. I was momentarily fearful that I wouldn't be able to match the right mark with the right sound to produce the right word, but then something flowed into me which calmed my anxiety. I felt an assuredness that one day I would know how to match the right sound with the right mark, that I would know how to read, and that was sufficient for me. Precisely as it was when I first encountered the technical terms utilised in the Work; in time I would understand.

Over the weeks and months that followed my introduction to the Work, life took on a different dimension and though many questions which had been in my mind for so long were finally answered, many more took shape. My mind turned inward such that the Work became the paradigm for conceptualising the other-worldly or non-ordinary experiences which had occurred throughout my life. I learnt that birth was a prelude to death and that the time in-between provided the opportunity to 'grow' my soul, transform my essence and work toward determining my spiritual destiny.

It sounded so deceptively simple, however when someone enters the Work they're required to engage with a particular kind of knowledge and particular kinds of purposeful activities. This not only challenges existing belief-systems, it also stretches the mind because repeated intentional effort of a particular type is required. This is necessary because we need to understand relevant concepts in order to be able to apply them to our lives.

I understood that life existed for a reason, that human beings were an intentional arising, and that creation was an act generated, brought forth and maintained by an almighty and immense love which permeated and sustained all existence. I saw that there was a cosmic order which maintained and supported the universe and that my life had to consist of uses. And although it took some time, I learnt that not only was I to be useful to my fellow beings, I had to be useful to God as well.

This was a concept that had never previously entered my mind. How is a person of use to God? What does one do? How could I, a tiny speck of insignificant organic life in the universe be of use to that which had lovingly created billions of life forms, had generated an infinite number of galaxies and had populated the universe with a myriad of planets and suns? My being born had incurred a cosmic debt that had to be paid, I understood that. I also understood that the abiding and loving vastness that had first communicated with me in my childhood had brought me forth; how was I to repay that?

It wasn't just my mind however that came alive as I digested the Work; it was my heart as well. I knew I had come into contact with something incredibly precious, with something so profoundly beyond what 'I' am, that I could never truly comprehend exactly what it was or how it had come to be. Yet, and despite the fact that I struggled to understand the Work, I *felt* its worth and something in me wanted to hold it carefully in my hands and keep it safe from harm.

The Sufi's say there are three journeys in life. The first is the journey to God. The second is the journey in God. And the third is the journey through God. And this is what constitutes the Work. Metaphorically, the Work can be contextualised as the ship in which we steer our spiritual course through life; it is a body of esoteric doctrine, method and instruction which enables us to journey to God, then journey in God and finally, to reside in God. That made sense to me.

In the early days, as I call them, I think I went around in a kind of euphoric stupor. I felt like a sponge soaking up not water but knowledge and truth that I knew would transform me as a human being. From the moment I entered the tavern and drank my first drink I knew that my life would forever consist of moments of drunkenness and sobriety, and that neither I nor my life would ever be the same again. I remember having endless discussions with other students. There were dinners and picnics which were occasions for us to come together outside the School where we could air our thoughts and muse upon our mystical mutterings, and even if we didn't understand what the other was saying it didn't matter. To be in the company of other like-minded people was for me an absolute joy because we all spoke the same language. Our minds connected each and every one of us intimately, so much so that it truly was a blissful period of my life.

After some months however, all manner of difficulties and problems arose. What was going on? Why was this happening? Was I not working on myself properly? Should I try harder? Was it my understanding? It took me a long time to realise that these happenings were events which were then, as they are now, challenges being presented to me by life. These events are necessary because they provide the raw material required for work-on-oneself, and to grow spiritually we have to be able to meet such challenges in a practical manner. When things no longer go rightly for us, it is because the expectations of our imagination are receiving a very big jolt and a lesson about life is being communicated to us. The following parable written in the language of correspondences captures exactly what was taking place at the time.

> And the voice which I heard out of Heaven, again speaking with me, and saying, Go, take the little scroll having been opened in the hand of the angel standing on the sea and on the land. And I went away toward the angel, saying to him, Give to me the little scroll.

> And he said to me, Take and eat up it, and it will embitter of you the belly, but in the mouth of you it will be sweet as honey. And I took the little scroll out of the hand of the angel and devoured it, and it was in the mouth of me as honey, sweet; and when I ate it, was made bitter the belly of me. (John 10:8-10).

The language of correspondences, which Emanuel Swedenborg introduced and expounded in his spiritual writings, instructs that spiritual things both reside in and are represented by natural or material forms, which themselves are just an appearance. "Correspondences are powerful instruments that link the transpersonal with the personal and bring the dimensions of infinity and eternity into one's life" (Groves, 2014, p. 183). Concerning parables, Fourth Way teacher Dr Maurice Nicoll observes, "All sacred writings contain an outer and an inner meaning. Behind the literal words lies another range of meaning, another form of knowledge" (Nicoll, 1984, p. 1).

How are we to understand these writings and how are we to establish a functional, psychospiritual relationship with them? The Work tells us to apply the Law of Otherwise, which operates on the principle that things are other than what they appear to be. This may sound simplistic, but the key is to remember the Law and then apply it. And we apply it by learning the language of correspondences, which takes time.

Regarding the parable of the scroll previously mentioned, the language of correspondences tells us that the scroll is a body of spiritual doctrine, which is the Work. The sweetness corresponds to the delight we feel when we initially encounter this body of doctrine. But there is a price to pay, which is the bitterness experienced by the belly. The belly corresponds to what is inmost within us, within the inner self. When we encounter the Work, it is not unusual to find that instead of situations running more smoothly they get worse.

This is because our negative states are stirred up and rise to the surface. The action of these states produces the resulting bitterness; the truth of a situation is sometimes hard to stomach. And this is exactly what occurred to me.

The action of the Work in our lives strips away our illusions of ourselves as it does our incorrect view of reality. It must do this so we can see our own inner reality as a spiritual being which is living in a physical or material body, as well as the reality of the spiritual universe. Working with our negative emotions is hard work. Accepting responsibility for ourselves and our actions is challenging. Re-educating ourselves in the light of sound spiritual knowledge takes time because ultimately, it is the mind that has to be renewed and transformed.

I recall when I commenced work on the manuscript for this book I had been pondering on happiness. Why wasn't I happy? Why had things gone wrong yet again? Of course, and though I didn't realise it, I was singing a repeating song and playing the same old record. I also recall at the time synchronistically chancing across Volume 4 of Dr Maurice Nicoll's *Psychological Commentaries on the Teaching of Gurdjieff and Ouspensky* (Samuel Weiser, Inc., 1996), and idly opening the book to a random page, was confronted by the following words:

> Why do some of you think that what matters is whether you are what you call "happy" or not? ... It is not that the Work is to make you unhappy, but that work on yourself will change your idea of what it is to be happy ... *This Work is to make something in you independent of what happens to you* ... How can you think that inner peace depends on what happens to you? A machine will behave in that way but not a man or woman who is trying to become more conscious and tries to meet all daily events, whatever may be their complexion.

> Therefore you see the Work is about making some point in yourself more or less independent of the happenings of daily life, and the apparatus of this Work and all that it teaches you about Self-Observation, Self-Remembering, and so on, has to do with making a point in you distinct from the machine that is governed by happenings ... (pp. 1426-1427).

And here I saw the crux of the matter quite clearly. The action of the Work is to make something in us independent of what happens to us. Referring back to the parable from John, the bitterness in the belly corresponds to our struggle against ourselves, our negative states and our reactions. It is our struggle toward transformation as it is our struggle to move up the scale of inner development and away from the Realm of Outer Darkness in which Man Number 1, 2 and 3 reside. Our efforts to live life in a more meaningful manner place us under different influences and propel us to the next stage of spiritual evolution whereby we transform to Man Number 4, 5, 6 and finally 7, which is Harmonised Man (see Chapter 5: Work Basics for additional information regarding the Scale of Gradation).

There have been many events and periods in my life that have been joyful and a delight to experience and none more so than these early days, as I call them. Since that time the Work has continued its action in my life, bringing me face-to-face with many of my emotional and psychological difficulties as well. There have been times of consuming psychospiritual darkness, as there have of mystical illumination. Through it all my faith and trust in God has both nurtured and sustained me. There is a lot to be accomplished in this life and bitterness in the belly is a small price to pay.

3

KNOW THYSELF

The conclusion that man is little more than a machine, controlled by external influences, asleep, and with virtually no power of choice, will no doubt appear to many as a singularly obnoxious form of pessimism. This is aggravated by the assertion that the very nature of time is to be the condition of decay and dissolution, thus denying any possible validity to theories of mechanistic evolutionary progress. It is not pessimism to discover that the things we seek in the future are not to be found there, and the pride that we take in human achievements is an illusory and damnable pride. In discarding our illusions we discard nothing that has true value, but, as it has been my main object to show, we open the door to a reality which far surpasses the utmost that mere human hope can dare to conceive.

John G. Bennett, *The Crisis in Human Affairs*, p. 14, 1954.

The 4th Century B.C. Temple of Apollo, the remains of which are located on the southern slopes of Mount Parnassos at Delphi in Greece, contained three instructions inscribed on the lintel which rested above its entrance. The first was, *Know thyself*, the second was, *Nothing in excess* and the third, which came from the Plato Academy read, *Let none enter here who cannot do geometry*. This book in its entirety aligns itself with the first instruction; *Know thyself,* which is what the Work requires us to do.

English Anglo-Catholic author Evelyn Underhill (1875-1941) observes that when the self begins to stir and to question its existence it is accompanied by a change of attitude, and perhaps a restlessness too. This attitudinal change drives the self to turn away from what Underhill describes as the unreal world of sense in which it lives and in which it is incapable of what she defines as super-sensual adventure. The result is that the self embarks on a journey toward God because there is simply no recourse.

> In the moment of its awakening, it is abruptly made aware of this disability. It knows itself finite. It now aspires to the infinite ... Whatever form, then, the mystical adventure may take, it must begin with a change in the attitude of the subject, a change which will introduce it into the order of Reality, and enable it to set up permanent relations with an Object which is not normally part of its universe ... We see a sham world because we live a sham life. We do not know ourselves; hence do not know the true character of our senses and instincts; hence attribute wrong values to their suggestions and declarations concerning our relation to the external world. That world, which we have distorted by identifying it with our own self-regarding arrangements of its elements, has got to reassume for us the character of Reality, of God. (Underhill, 1995, p. 199).

Underhill observes that because of not knowing the reality of ourselves we correspondingly don't know the spiritual reality of the world in which we live. To address this, we need information and instruction. We need to know *how* to know ourselves, and we need a special *kind* of knowledge which in addition to being truthful and practical, needs to have been tried and tested by others. We also need the assurance that we can trust this knowledge to guide us toward the understanding that is necessary.

In addition to knowing ourselves however, we're also required to understand the spiritual framework which provides the foundation for the material universe in which we live. To do so, again, a special kind of knowledge is needed to help us both understand our place and purpose in the cosmos, and which can unite us with Higher Life. Here the Work provides practical guidance. What is ancient is best. This knowledge has been tried and tested by countless people throughout the ages. It is eternal, enduring and rooted in truth. It has been kept alive in closed and hidden esoteric Schools in the East and the West for thousands of years, and it has endured for a reason.

Finding expression in verbal and written transmission, objective art, music and dance, the Work seeks to educate people so they can awaken from the sleep-state. Being ignorant of this reality is not an impediment precisely because being ignorant provides an opportunity for us to become informed, to become educated. Those engaged in the Work are spiritually known as 'the lungs of mankind' because their combined efforts uplift humanity in various ways. It is through these activities that knowledge and information regarding the inner reality of the universe can be transmitted into receptive minds.

What is the Work? The Work is a body of esoteric knowledge, doctrine, and instruction. It has been kept alive during the history and ages of mankind in various centres of instruction known as Schools. Scattered on the planet's surface and around the globe, these Schools disseminate the knowledge and instruction of the

Work to students who attend. To a large extent the knowledge and teachings are hidden, simply meaning that the Schools which house the knowledge are not to be found alongside mainstream teaching institutions. That is not to say that the knowledge is exclusive, that it is kept away from others, or that people are denied access to it. Quite the contrary; everybody is granted that which they seek. Everybody will find through searching that which is most fulfilling and meaningful for them. The truth of the matter is that most people are either not interested in the Work or find it too difficult to embrace.

People generally have an incomplete view and understanding of the spiritual reality of the universe in which they live. This is due to the ongoing impact of the sleep-state and an abundance of fantastical theories, fancies, and notions about life. Esoteric work can address this because it brings people back into participating in the wholeness of reality.

We live in an age, particularly in Western or first world cultures and societies, where the need and demand for new and ever-changing technology runs the risk of overshadowing a religious or spiritual way of life. An important consideration concerning this is that each successive generation is at risk of becoming more materialistic than its predecessor, and people everywhere, from a Work perspective, seem to be forgetting more and more why they are alive. Most people, as repugnant as it may sound, are firmly enmeshed in external life, are sleeping, sensory-bound and do not know what they are living for. And for many, one of the most fundamental questions of life, 'Why was I born?' never enters their mind.

We need to know what reality is and how to actively participate within the framework of that reality. All of us are born in innocence and ignorance and need to be instructed in how to live spiritually, and it is the Work which can provide that instruction. The Work is how we can procure for ourselves an eternal conscious soul. However, doing so requires effort and the deliberate abandonment of one way of living in the world, for another. The aim of the Work is very specific; it seeks to arouse and educate the mind to awaken it from the sleep-state. But there is much more to the Work than human beings simply awakening from their sleep-state.

The Work is not a quick fix for people seeking overnight enlightenment. It has nothing to do with channelling, contacting spirits or spirit guides, or developing magical or healing powers and abilities with which to influence or control others. It does not teach the doctrine of reincarnation or karma. What it is, is the science of self-knowledge, self-development, self-cleansing and rebirth. It is a constant activity and discovery of what we are and of what we are living in the midst of. It is concerned with the attainment of consciousness and unity, the growth of the soul and the working toward of our spiritual destiny. The application of the Work in our lives restructures the mind and reorganises the life, an activity termed the *Second Education*. Second Education is a technical term and has a distinctly different meaning to the first education people generally receive.

The Second Education, being our inner spiritual education, differs to our first education which is orientated toward building the personality (which is the outermost part of the psyche) and to relating ourselves in a functional and practical manner to the world and society in which we live. Like the first education, the Second Education is concerned with aims and goals. The difference between the two however, is that the first education equips us with the knowledge and means to live and experience our embodied lives, obtain an education and livelihood for ourselves. The Second Education is completely different because it involves seeking and having access to esoteric knowledge for the purpose of soul growth and vivification of the essence. This education is orientated toward self-transformation, spiritual regeneration, and growth of being. Love for God and recognition and interaction with Higher Life are primary components of our Second Education, as is growing toward the understanding of what we are and what we can become. It is important to remember that when we begin our Second Education, we embark on a career path with eternity in mind.

> Our second or inner education completes the formation and perfecting of our whole life, renders our internal being conscious and active, and helps us to understand the nature and destiny of our existence. Finally, it leads to harmonious relationships between the inner self and the outer self, in which the enlightened inner self guides and inspires the outer self with love and wisdom, and the outer self feeds the inner with the energies and experiences it requires and provides for the expression of its qualities and emanations. (Groves, 1996b, p. 1).

The Work stresses the necessity of relating oneself to God and binding oneself back to the creative source. It is not merely an intellectual endeavour, nor is it given to those who seek it to be casually used in armchair discussions. When we enter the Work, we are committing ourselves to an eternal relationship with God and Higher Life, and we need help to do that; of ourselves we can do nothing but with God all things are possible.

It is helpful to see the Work as a pathway of psychospiritual development, instruction, education and once engaged with, growth. It has been given to mankind and made available to all by Higher Life with the aim and intention of regenerating and restructuring the human psyche in preparation for entry into the spiritual universe. In developing a relationship with Higher Life, the Work necessitates that not only do we need to know ourselves, we also need to develop an understanding of the functioning of spiritual reality. In knowing ourselves, the ofttimes painful truth of revelation is followed by the birth of understanding and the knowing that change, if we want it, is a certainty.

The material universe in which we live as embodied beings has in Indian philosophy been termed the world of 'Maya' (Vroom, 1996, p. 57). This is the Hindu concept of illusion, and it warns that we are deceived by our senses into believing what is real, particularly regarding the reality of the spiritual universe. The world of Maya, the world in which we live as embodied beings, is a place where negative emotions dominate, where there is ignorance of spiritual realities and where there is much misunderstanding and suffering; it is the world of self or personality.

I remember a Work exercise which was designed to illustrate the precarious nature of living in the world. Students were instructed to go to the seashore at low tide when the rocks were exposed and still wet and slippery. We were to walk carefully over the rocks taking care not to slip, so avoiding possible injury, as we were to watch for freak waves, so avoiding possible drowning. "The world is a dangerous place" we were told, "and this is how you must walk through life, carefully, and on watch for the threat of ever-present danger" (*Author's journal notes, 17 May 1989)*.

The Divine loves and cares for the welfare of all individuals and does not sit in judgement on any human being, dispensing fortune here and calamity there; God does not have favourites. All of mankind is loved equally, simply because the Divine loves all that it brings forth from Itself. Events frequently occur over which people have no control, that happens, however it is human beings themselves who generate their own happiness or unhappiness by the attitudes they cling to and by the behaviours they adopt in response to those events; people everywhere co-create their present moments.

And as far as holding God responsible for what occurs in the world and in our lives, what we need to understand is that it is *life* that is challenging us, not God.

The Divine generates spiritual essences which are manifest in human beings and born onto the surface of the planet. Here we endure various types of suffering and hardship which provide the conditions for work-on-oneself to be carried out. The society into which any essence is born is an environment created by those that constitute the society with their political, civil, and moral codes of law, ethics, and conduct. If there is political, civil, and economic unrest or political, civil, and economic instability then we need to consider that those conditions at some moment in time ultimately have their origin in human choice, decision and action. This has nothing to do with God and everything to do with humanity which in its decision-making, either just or unjust, creates its own structured societies, and social, ethical, and moral morays and dilemmas.

The Divine, in generating an essence full of the potentials of spiritual growth and development, has provided each and every one of us with an enormous head start in life and it is our choice whether or not to actualise and maximise those potentials. It must be noted that in some cases unexpected death early in life may inhibit or interfere with the possibility of spiritual growth and development. We can choose to develop into the fullness of a conscious being, or we can ignore the invitation given to us by Higher Life and consign ourselves instead to a life of spiritual sleep. The spiritual universe is an active place, and we need to ask ourselves whether we are going to participate in the fullness of that activity. The Second Education requires that we make something of our lives which will endure beyond physical death; this is the urgent and pressing task for each one of us.

When we die, we take with us what we are at that moment in

time, that is, what we have made of ourselves up to that moment and no more and no less. If at death we are still in the grip of the sleep-state, then we take that with us into the spiritual universe. Do we want to help the spiritual universe grow, as it does through the acquisition of conscious souls who form spiritual and celestial communities and societies living lives of use and contribution, or do we want to spend eternity shrouded in spiritual darkness and ignorance, dead to the joys of living? The rule is this, as we live here so we live there. We create our spiritual destiny as embodied beings, and we must work on ourselves with a sense of urgency in order to do so.

Death is the putting off of the physical body and is the event which enables us to enter the spiritual universe as a disembodied or spiritual being. There is a constant stream of disembodied beings entering the spiritual universe at any one time, with the World of Spirits their first destination. It is here that, amongst other things, newcomers from earth are welcomed and old acquaintances renewed. Some stay momentarily before moving on to their spiritual community in other locations of the spiritual universe, where they live eternal and joyful lives of use. Others linger for extended periods and sometimes develop close associations with still embodied planetary dwellers, at times impressing them with their own thoughts, memories, and desires. This can occur to such an extent that the previous earthly life lived by the disembodied spiritual being is considered a previous karmic life of the still embodied planetary dweller. The World of Spirits is the dwelling place for those disembodied beings that identify themselves as spiritual guides to spiritual mediums.

Most spiritual beings in the World of Spirits are considered low-grade spiritual beings. This is not a criticism, far from it; it is simply the reality of the situation. They are like this because their own actions while living as embodied beings consigned them to that realm. This is their spiritual destiny; this is their eternity. In his five volumes *Spiritual Diary* (THE SWEDENBORG SOCIETY, 1883-1902/1977), Emanuel Swedenborg provides an extensive and meticulous record of all manner of diverse interactions with spiritual beings from the World of Spirits, including with newcomers to that realm as well as those who have dwelt there for varying periods.

For the most part, human beings don't understand or fully comprehend that an enormous amount has been invested in them, and that in truth we are all cosmically indebted to God for our arising. How is this debt to be repaid? We work on ourselves, we rid ourselves of our negative states and reactions, we pave the way for the entry of higher truth into our being and we undergo transformation so we can become of use to God and to our fellow human beings. Additionally, we must earnestly strive for the light of consciousness to flood our inner being so it can bring to the fore more deeply buried material that must be worked upon.

The Divine loves its creation and wishes to bring all human beings to their greatest happiness; however, it is human beings themselves who block the entry of Higher Life and who mar their own spiritual happiness. This is where the Work can assist us because it provides instructions for how to engage with Higher Life in a practical way. It is a Work teaching that spiritual enlightenment is only granted to those who are spiritual. At first glance this may appear harsh and judgmental, however many people live in the lower part of the soul and mind, a situation which makes it impossible for any form of spiritual enlightenment to take place. But human beings can choose to make decisions for themselves, and in doing so bear the responsibility and consequence of those decisions. Enlightenment can occur, but we have to want it.

As well as having a powerful orientation toward Higher Life, it is important that we be realistic about our role in life. We have to develop the best in ourselves without loving ourselves. This is a subtle point because we have to eliminate 'love-of-self' which can take many forms, including vanity. Vanity becomes egotism, manifesting itself in our thoughts by way of thinking that we are better than others. We look down on them in a condescending manner and dismiss them somewhat high-handedly as 'sleeping people'. Vanity induces us to think we are full of wonderful things. We are not full of wonderful things. Honest self-observation reveals quite clearly what we are; masses of negativity, a disunited whole, and an unconscious machine which simply reacts to occurring events.

> What is desperately needed is a most profound awakening of the human mind from all the narrow ideas, stupidities, theories and dark dreams that block up the delicate passages of the spirit. What is needed is a working vision of an infinity of greater realms extending on all sides and also here in our midst at this very instant. What is needed is a relentless casting aside of the gravitative "I" which pulls an endless assortment of heavy energies towards itself by its ever-unsatisfied wants. (Groves, 1996c, p. 100).

If we are to reconstruct a new life for ourselves, if we are to have a 'working vision', then we must remove our impediments. This is hampered by the fact that we carry within ourselves a barrier which hinders us continually. The barrier is composed of various things including egotism, vanity, pride, negative emotions, and unjustified assumptions. We all have this internal barrier which causes us to live in the outer or external part of ourselves, the personality. This barrier produces the sleep of mankind; the sleep-state. In this situation it is negative emotions which rule, and which consequently produce problems, worries and difficulties. The solution is to introduce spiritual knowledge into this external part of ourselves which can be cherished and loved. We must compel ourselves to make this knowledge organic so we can live effectively. Quite simply, we build it into our lives and we live from it.

This knowledge is often referred to as the 'pearl of great price'. The following parable refers to the pearl of great price whereby a man gives up all his former knowledge because he has found a greater knowledge, he has found esoteric knowledge.

> Again, the kingdom of Heaven is compared to a man, a merchant seeking excellent pearls; who finding one very precious pearl, going away has sold all things, as many as he had, and bought it. (Matthew 13:45-56).

We must transform the outer part of ourselves and erode the barrier within ourselves. Esoteric knowledge alone provides us with the means and know-how with which to achieve this. This may take years but if we work on ourselves the barrier will be breached. This will enable the full flow of life into us in an uninhibited and unobstructed way, after which something from above can transform our outer life. Esoteric work insists that knowledge and being must keep pace with each other. What we know and what we will, needs to be built into our being so that it becomes the power of understanding.

When we die our external knowledge disappears and it is our being that gives us entry into a particular spiritual society. What we are is our being. Knowledge is a property of personality, but understanding is a function of being. When we work on ourselves, we are looking at connecting ourselves with the wholeness of reality and this is the aim of Beyond the Fourth Way. We are created to be channels through which an endless creativity can and should pour through, and we can ensure this by working to remove the barrier which resides within. Every person senses a state of isolation at some time or point in their lives. This is overcome when we break through the barrier which in turn allows us to sense and feel the flow of Divine life within ourselves.

To work-on-oneself means many things. In some schools the following three-fold principle applies. We must first be Deists then Humanists and then Naturalists. A Deist is someone who sees God and refers to those who have a genuine love of God. This is an inner relationship between that person and the Divine which is arrived at after much pondering and reflection. The Humanist is someone who perceives the whole range of human life and history and loves the pattern that is within all human beings which can develop endlessly within a person. It is to love the essence of the essential human being. The Naturalist loves that which stands behind creation. This person sees that behind everything that exists there is a correspondence that indicates a Divine, celestial, or spiritual function.

Rather than just being only one, two or three, our task is to be all three simultaneously; we must link and connect the Deistic, Naturalistic and Humanistic function within ourselves so that each illuminates the other. When this happens the whole of our being becomes unified. This is possible because of the indwelling presence of the Divine, which is something we have to gradually experience. We must remember that the indwelling Divine presence is constantly with us irrespective of whether we believe it or not or know it or not. At all times and in all situations, the Divine resides within us, and we must make ourselves aware of this fact. In the chapter titled *Intentional Effort,* I include an account of an experience which demonstrates this.

When we live the Work, we begin to understand it. But it's more than this. It's loving and cherishing it and willing it into our lives, this is the crucial element. The ability to embrace the meaning of the phrase, 'the Work is God's outreach to humanity' arises from a shift in how we conceptualise our relationship with religion and spirituality, how we live in the world, and in knowing and understanding the purpose for why we were born. The shift produces an internal inversion, a psychospiritual transformation of our inner life, and once that occurs, and once we understand the significance of what has happened, nothing is ever the same again; in a sense we are freed from the confines of the external self.

Returning to the phrase, 'the Work is God's outreach to humanity', we need to embrace that outreach with all our heart, revere it and make a place for it within ourselves where the knowledge is kept safely tucked away from the outside world. We need to create for ourselves a Temenos, a sacred space within ourselves where we receive the influx of the Divine and concentrate it so that it is always there. Once this has been created, we have to nourish it by thinking spiritually, which is different to ordinary sense-based thinking. For example, we concentrate on collecting spiritual truths and we study sacred writings such as those found in the Bible, the Gnostic texts, the Koran or the Bhagavad Gita. In doing this we are trying to both move beyond our veiled understanding of reality and to consciously connect ourselves with a bigger and more expansive idea of what reality really is.

Work-on-oneself takes time. Work ideas and teachings need to be gone over time and time again and it may take months or years to grasp and understand even elementary principles; what is important is that we keep at it. The duration of our earthly or planetary life is ample time in which to build for ourselves a Kesdjan or spiritual body, and in which to complete our Second Education. When we finally enter the spiritual universe to claim our rightful heritage in eternity, it will be, if we have worked on ourselves sincerely and diligently, as loving, useful, and functional components of the Grand Man, Maximus Homo.

> Man can never be MAN as long as he merely wishes, wants and desires that which is insignificant. In his despicable darkness of mind he half wonders whether there is life on other planets, or whether there is really life beyond this physical body, but he cannot really bring himself to believe. He wants proof; he wants to see with eyes that are only fit to look upon shadowy reflected shapes that it pleases him to call reality. (Groves, 1996c, p. 100).

Beyond the Fourth Way and the Fourth Way combine to form a powerful means by which we can learn and understand the spiritual reality of ourselves and the material universe in which we live. Born in love, the Work *never* imposes itself on human beings; no person should be compelled to do anything in life, much less work on themselves. Perhaps we can understand the Work as the manifestation of the Divine Love and Wisdom operating in a manner accessible to us all. To access that, what we need to do is knock on the door which will grant us entry to Higher Life. And if the door doesn't open on the first knock, we knock again. If it doesn't open on the second knock, we knock again. And if we knock loud enough and long enough, Higher Life will respond, and the door will open.

4

TECHNICAL TERMS

> All sacred writings contain an outer and an inner meaning. Behind the literal words lie another range of meaning, another form of knowledge.
>
> Maurice Nicoll, *The New Man*, p. 1, 1984.

The Work contains many technical terms which are important to understand. These terms form the backbone of esoteric knowledge, and they must pass from the memory of the external mind into the understanding of the essence. In carrying this out, what we are endeavouring to do is to consciously redirect esoteric knowledge to the essence which requires it to grow. Truths must pass from the personality into the essence, and it is our duty to see that they do.

The vivification of the essence is crucial to determining our existence as a spiritual being in the spiritual universe, for from the essence comes forth the soul. Higher Life needs to be fed with good quality souls because it is these which the spiritual universe uses to enrich and extend itself. To be of use to the Divine and our fellow human beings, it is imperative that as self-evolving organisms we develop ourselves to the highest possible degree. This is what the Second Education is all about.

To convey as accurate a meaning as possible of the following selected terms, I have drawn largely upon my notes taken during various lectures I attended spanning the duration of my time as a student at the School. For additional insights however, I recommend to the reader Maurice Nicoll's book, *Simple Explanation of Work Ideas* (Eureka Editions, 1998), which provides a fuller explanation of Work ideas in jargon-free English.

ARCHETYPES

Being the one race, the human race, human beings share a collective unconscious that links them at a common root. The collective unconscious is that which conjoins all human beings at a deeper phylogenetic spiritual level, like a vast pool or reservoir feeding directly into every individual. It is into this collective unconscious that the Divine injects archetypes, spiritual patterns, which criss-cross and permeate all cultures. Archetypes have a spiritual source and are living and dynamic symbols of higher realities. As a result of this common pool which is called the collective unconscious, the individual has within themselves many levels of archetypes including celestial archetypes, spiritual archetypes, sub-human archetypes, animal archetypes and plant archetypes. These levels are present within all human beings because they are basic universal patterns which are built into us, and which are the source for much of the symbolism from which dreams are composed.

BEING

All that a human being is within themselves, their thoughts and feelings, their unconscious, the sensitivities of their responses to the universe, this is what makes up our being. On a small scale, we are trying to come to terms with our being and it is through observing, sensing and feeling, that we gain an intimate awareness of our being. In its simplest and most direct form, we can say that being is you, it is what you are, what I am, what we are, and to have being is to be alive, to exist. In essence, 'being' is an all-encompassing term for who and what we are as a person, as an individual. It is everything that makes us, 'us', that makes me, 'me' and that makes you, the reader, 'you'. The affirmation 'I am' is a declaration that "I exist as a living being".

CHIEF FEATURE

Associated with the proprium, otherwise known as False I, is the chief feature. This is our greatest weakness because it works against us by exerting a tremendous influence on the outer self. Chief feature is a negative attribute that must be changed and modified and made into a more heavenly thing. Chief feature in its own way can be a gem because it prompts us to make efforts to overcome it, and it is these efforts which have a positive effect. Chief feature acts as a constant stimulus, encouraging us to try all sorts of things to overcome it; it drives us to find positive and useful ways in which to better ourselves. Negative emotions express certain aspects of the chief feature. It is not their force but their persistence which is the diagnostic tool. It is important to notice how often these little negative states arise, that is the key.

CHURCH

The term church represents a particular orientation of mind in which there is an intimate, loving, and functional relationship between the Divine and the individual. When we work upon ourselves, we in effect become a church in its real sense because we endeavour to pave the way for the reception of influences coming from Higher Life. We know and experience the reality of our cosmic situation, our ignorance and lack of understanding, our helplessness, and our complete dependence on God. Church is the internal process of developing conjunction with the Divine. The Work seeks to transmute our mind from a sleeping to a wakeful state, as it also strives to bring the heart into a state of loving remembrance of God. When these are active within us, then we are said to be a church.

CONSCIOUSNESS

It can be difficult understanding what consciousness is because people generally confuse it with awareness. Many people believe they are conscious individuals, however in the context of the Work they are not conscious. What they are working with and experiencing is the energy of awareness, which is a lower grade of energy than that of consciousness. It is this awareness which must be transformed into consciousness, which is a spiritual light originating from the Divine. Without becoming conscious, an individual can never break free from the darkness which encloses their inner life.

Consciousness may often arise as a subtle perception, perhaps as a gentle noticing of things in our self or the environment never noticed before, or it may be insight into a meaning that previously avoided us. Loosely translated, consciousness is the knowing of things in a state of togetherness. It is not an energy normally found in the body or nervous system but rather is a kind of inner light which illuminates and makes clear what was hitherto not clear or not seen. Because it lacks the power to do anything, consciousness can only behave or perform in this way.

In the human psyche there is a lot of darkness. When we become conscious, energy, as light, pours into the psyche dispersing the darkness and enabling us to see what is there, but that is all that it does. Without consciousness nobody can see where their difficulties lie, which is why our first great effort in life is to become conscious. Consciousness by itself is useless unless it has a content, so it must be accompanied by practical truths and teachings. We must *know* what to do with the things that we see which need to be put into order, and this is where the Work comes into play.

EGO

When a human being is born, they contain within themselves all archetypal patterns of wholeness and are 'real Self. Self, while being the embodiment of the Divine, builds a structure which can fend for itself in the world and in turn manufactures the ego. The Self is born but the ego is made; this is an important distinction. The ego is the instrument which enables us to relate to, and interact with, the world in which we live; the visible, material universe. The ego is also the centre of consciousness, and it is a necessary instrument for forming a link between the self and the outer world. However, the formation of the ego leads to a partition in life. The ego must be differentiated from the self, a process that generally occurs during the latter half of life (which is when the reintegration of this process should occur).

Careful and conscious examination of oneself shows that the ego is a cluster of states which include psychological states, emotional states and physiological states all held together by habits and chronic feelings such as self-regard and self-love. Part of esoteric work on oneself is to see the non-substance of the ego and to realise how flimsy it really is. If these habits and chronic feelings can be overcome, we make room in ourselves for the entry of our individuality which is given to us from Higher Life, an event which constitutes a rebirth.

ESSENCE

Great emphasis is placed in the Work on the importance of arousing the essence. The essence, which is a human being's spiritual component, contains within itself fine substances and energies which hold all the potential for the eventual development of a human soul. In order for it to come forth from its internal abode and assume its rightful role in our lives, essence needs the help and assistance of the personality. It is the duty of the personality to reach the essence, arouse it and then feed it with truth so it can become a fully functional soul.

Whereas the personality accumulates knowledge, which is always about something, essence needs understanding. Things will only pass into the essence when there is understanding. It is imperative that the Work passes through the personality and into the essence, and a primary aim of the Work is to teach the personality how to reach the essence and vivify it. If we are to grow a soul for ourselves it is vital that this truth be understood and willed into action.

FALSE-PERSONALITY

False personality is everything an individual assumes they are but in truth are not; it is something that arises out of erroneous assumptions regarding ourselves. False personality is a smoke screen which we generate; we make ourselves smarter and better than other people. To deal effectively with false personality, and to pave the way for the personality to assume its correct role in our lives, we must eradicate it. This is done by refraining from lying to ourselves and to others, by not identifying with negative emotions and reactions and importantly, by avoiding theories and assumptions. The false personality is not 'us' yet for most of our lives we live from this part ourselves. False personality is the root cause of many of the difficulties we experience in life.

LEADING LOVE

When we are alone, our leading love looks out and fashions our line of thought. It is important to find time to be alone because when we do so, we can observe these lines of thought, which in turn enable us to determine our leading love. The leading love shapes the will, which may not necessarily be just one thing and it may not necessarily be a negative thing. It may be for example a person's natural drift in life, or it may be geared toward selfish needs, wants and pursuits. It may have associated with itself secondary loves such as self-importance or the desire to obtain numerous degrees or diplomas to make the individual appear 'important'. The leading love is self-seeking. It is what we secretly yearn for, and it subsequently shapes our whole life. The only way to detect the leading love is by very long self-observation. If it is discovered that the leading love is perverse it must be put right.

MAGNETIC CENTRE

Magnetic Centre, one of the technical terms utilised in the Work, is a slight extension of the spiritual mind into the realm of the natural mind. If we are brought up as a young child with the belief that there is a great God who supports and sustains all things, such ideas coming from the outside will touch Magnetic Centre within us, causing it to enlarge a little. This is all that most of us need because for the next fifteen or twenty years the First Education takes place and the personality grows. By the time we reach thirty or so we should have made some sort of success of our life and being reasonably well established, now have the time (all being well) to extend our lives in an inner way.

If we start to look at things in a deeper way, Magnetic Centre will be ignited and after a while we begin to think and act in a spiritual way. Material security is important, but those of us whom see only this aspect of our lives as important are asleep. When we feel that there is more to life, that there is something beyond the surface, this is Magnetic Centre in action. Magnetic Centre operates in us as an attraction toward the spiritual, toward what is hidden. In the West we always assume that this lies in the East but in fact the reverse is true. This knowledge is here with us, however it's in a form that we often don't recognise. Once Magnetic Centre contacts the Work it dissolves because it has served its purpose, it has led us to where we need to be.

PERSONALITY

When a human being is born, they are born with an essence, a physical body, a blood group, an ability to learn, to breathe and so on. From the moment of birth, a diverse range of acquired materials such as behaviours, attitudes and beliefs are built around the essence and grafted onto it. As time passes, essence is largely ignored while instead this grafted material, which constitutes the personality, gets bigger and fatter. It is the task of the personality to gather information and experiences from the external world which can then be distilled and used by the essence. This rarely takes place however because when personality becomes strong it thinks that *it* is the life of the individual. Rather than serving the needs of the essence, which is its proper purpose and function, the personality instead generates what is known as the false personality.

REAL I

Real I is a touch of new life coming into us from a higher level of reality; this is rebirth. In life, the majority of people fail to recognise that they have an outer self and an inner self, which is deeply buried inside their being. The inner self and outer self are not only often in conflict, neither of them is the seat of Real I. The domain of the outer self is worldly, whereas the domain of the inner self is private. For the most part, people oscillate between these two worlds. Only by learning about False I can we see through it. Once we see through it, we then make room for a third element to make ingression into us, Real I, which is a touch of the Divine.

REFORMATION

Reformation occurs when the internals of an individual are being spiritually regenerated. If we can get into a better pattern within ourselves, everything will change. If we can find thought patterns, or a religious theme which has the right sort of shape or fit for us, it can be a new beginning of life. Human beings can only pattern themselves on that which is above or beyond them, on that which is spiritually stable and enduring. All growth takes place in discontinuous steps because there are stages of change. Because of our multiplicity we cannot become complete beings over-night, and because there will always be a resistance to whatever is done, we must allow for these stages to occur and for change to take place in its own time.

REGENERATION/REBIRTH

Regeneration or Rebirth represents the death of the old self and its erroneous and mistaken beliefs and attitudes which have to die, and the introduction of new and deeper knowledge beyond the old information. Rebirth, which is a fresh start at a deeper level, is a new outlook. When we undergo regeneration, we begin to catch glimpses of higher realities.

Associated with regeneration and rebirth is self-cleansing. This is the process of getting rid of resentments, hostilities, antagonism – we have to work hard upon ourselves to get rid of these things because we are usually quite attached to them. Regeneration of the person as a totality requires that the inside of ourselves is fed new truths and principles which we learn to love. Once the inner self begins to love these things it instructs the outer self. This is an action which is strongly resented by the outer self and one which inevitably results in conflict. If the inner self gains strength from the Divine, then the outer self becomes the recipient of what comes from the Divine and a gradual refinement occurs. When the inner and the outer-self agree, an integration occurs resulting in a spiritual life vastly different to the physical life we have previously been accustomed to.

SOUL

According to esoteric teachings, the soul is formed of the things of love and wisdom given by God. Life, love and wisdom flow in to sustain the pattern of the soul. Now the soul also needs an instrument of knowing and is given the power to generate this instrument. This instrument of knowing is a spiritual formation, the mind. The mind is capable of knowing the action of life and the impact of wisdom. It can feel the urge of the soul passing through it, something that becomes the individual's natural drift. The mind is an individual's most important function because this is the part of us which looks up or down. It is the mind that is the instrument of growth and transformation. The destiny of the soul is wrapped up in the mind. People often say they 'have a soul'. This is incorrect, they should say, "I am a soul."

The soul is a human being's inward reality and a function within us which enables us to move toward the Divine. As such, it is an activity, a principle within us. It has a basic framework which enables the energies of life to flow through it, which then in turn maintain our being. Contrary to how people generally understand the soul, in the context of the Work the soul is an incomplete spiritual structure which needs to be grown and which must evolve. At every instance the soul is receiving the Theomertmalogos the insights from which it then translates into the subconscious. The soul experiences the need to reach God, it needs to find strength, comfort and support in something greater than itself. Soul is essence transformed into a fully conscious state in the true sense of the word.

THE CENTRES

The Centres in a human being consist of:

1. The Thinking Centre
2. The Feeling Centre
3. The Instinctive Centre
4. The Moving Centre
5. The Sex Centre

The Centres are designed to be channels through which the essence can both gather information and express itself. Instead, they are commonly used for other things which results in a kind of autonomous mind forming around them. This series of minds all work independently and don't understand one another because they speak different languages. The minds or 'brains' quite literally have a mind of their own and are very often indifferent to one another, a fact confirmed by self-observation. The Centres are functions of thought, feeling, instinctive activity, movement and sex, which do not have a specific anatomical location in the body.

In addition to these five Centres, there are two others:

6. The Higher Feeling Centre
7. The Higher Intellectual Centre

The Higher Feeling Centre and the Intellectual Centre are names for highly sensitive functions that enable the individual to receive influx from higher worlds. The Higher Feeling Centre corresponds to the celestial realm with its fundamental characteristic being Divine love. The Higher Intellectual Centre corresponds to the better levels of the spiritual realm with its fundamental characteristic being Divine wisdom.

THE FIRST EDUCATION

The First Education opens the external part of the mind, which is its primary function, and helps to form the personality. It enables us to relate in a functional way to the external world and to hopefully become a useful member of society. During the period in which our First Education takes place, we are provided with information, taught certain skills, and generally endeavour to attain some degree of competence which enables us to establish a career for ourselves and to provide for ourselves and/or our family. The First Education is concerned primarily with knowledge that relates to human beings in general and the social and cultural environments in which they live.

THE SECOND EDUCATION

The First Education enables us to live and function competently within society. The Second Education is completely different because it involves seeking and having access to esoteric knowledge for the purpose of soul growth and vivification of the essence. The Work constitutes the Second Education and is how the individual, working in conjunction with the Divine, procures for themselves an eternal conscious soul. When the Second Education is undertaken, the individual endeavours to build for themselves a better personality by getting rid of false personality, which is a graft and full of pretence.

The Second Education paves the way for the essence, which is given to a human being by God. The essence has to be aroused and brought forth into expression because its activity is crucial to the growth and evolution of the soul. Due to lack of interest in the Second Education, many people enter the spiritual universe in an undeveloped and incomplete state with poorly formed spiritual bodies. It is important to remember that when we begin our Second Education, we embark on a career path with eternity in mind.

THE SLEEP STATE

The sleep-state describes the situation of a human being who has no spiritual perception, and who is unaware of the spiritual reality of themselves and of the spiritual origin of the universe in which they live. The sleep-state, which has devastating consequences because of the resulting spiritual ignorance which occurs, afflicts the entire human race and gives this planet the spiritual name of what is termed, The Realm of Outer Darkness. It is from the sleep-state that mankind must awaken.

TRANSFORMATION

Human beings are a mixture of thoughts, memories, emotions, beliefs, attitudes, and diverse life experiences. Transformation is the transition from being a mixture such as this to becoming, and being, an integrated unity. Transformation is never automatic. It is the transition from being a mixture of states within ourselves to becoming a unified living Soul and is a long and ongoing process throughout our life.

In concluding this chapter, explanations for only several technical terms have been provided, and only those which specifically relate to those used throughout this book. It has been my observation that people tend to automatically reject that which is unfamiliar or that which is contrary to how they think or believe things should be. I want to encourage the reader to be aware of this tendency, and to try to ignore it.

I'm particularly aware of this tendency in myself because every single time, without fail, that I attempt to read Gurdjieff's Beelzebub's Tales to His Grandson (Arkana, 1999), I huff and puff to myself because I don't understand the technical terms or because I can't bear the long unbroken sentences or because irrespective of what page I happen to be reading I can't make any sense of the storyline. And this is the point; not only does the external mind need to be 'shocked' in this way by encountering what is unfamiliar, the unruly outer self, the personality, must not be permitted to decide how things are to be when that happens. It is the Deputy Steward who are in charge, and it is they who need to guide us as we work on ourselves (see Chapter 7 for a fuller explanation of the Deputy Steward).

We do need to be challenged, we do need to be stirred and woken up, that is for certain. Work-on-oneself is not only a life-long activity, but also a constant dynamic engagement between ourselves and Higher Life. Overcoming the habitual reaction of immediately dismissing what we don't understand or what is unfamiliar to us is vital if we want to facilitate new trains of thought in the mind and if we want to grow in a new direction.

5

WORK BASICS

> In contrast to the mechanistic Cartesian view of the world, the world view emerging from modern physics can be characterized by words like organic, holistic, and ecological. It might also be called a systems view, in the sense of general systems theory. The universe is no longer seen as a machine, made up of a multitude of objects, but has to be pictured as one indivisible, dynamic whole whose parts are essentially interrelated and can be understood only as a pattern of a cosmic process.
>
> Fritjof Capra, *The Turning Point: Science, Society and the Rising Culture*, p. 66, 1983.

The Work attempts to instruct us with two lines or phases of development; the first is psychological, the second, cosmological. The psychological line of development deals with the cleansing and purification of the psyche and the integration of the individual into a unified and harmonised state of being. The cosmological line of development deals with the instruction of cosmic laws such as the Law of World Building and the Law of World Maintenance, the Scale of Octaves and the Ray of Creation.

In accordance with the psychological line of development, School work emphasised 'the basics'; rudimentary principles for work-on-oneself which included such activities as self-observation, non-identification and self-remembering. These principles are essential components of the Work, and as students we were expected to understand and apply them to our lives. They have an integral relationship with our being-growth and development, and I recall that I found it helpful to position that relationship within a context of 'types of being'.

These types of being are conceptualised as a Scale of Graduation in the following manner:

Man Number 7	Perfected man
Man Number 6	Celestial man
Man Number 5	Spiritual man undergoing transformation
Man Number 4	Harmonised man
Man Number 3*	Intellectual man
Man Number 2*	Emotional man
Man Number 1*	Man of action

[* Man Number 1, 2 & 3 all stand on one level.]

The Work teaches that in accordance with this scale, all of us are either Man Number 1, 2 or 3. Once we have begun work on ourselves however, we can alter our positioning by growing and developing beyond each level. Man Number 1, 2 and 3 all stand on the one level. At this level the three Centres, Intellectual, Emotional and Acting operate independently. Man Number 4 is different because he knows how to link the Centres so that they act in harmony together. At this level the higher Centres open up within him and there is a corresponding influx of spiritual energies coming into his life which precipitates a range of entirely new experiences. If Man Number 4 decides to stay where he is, he may slide back to

Man Number 3. If he decides to go forward however, he will in turn become Man Number 5.

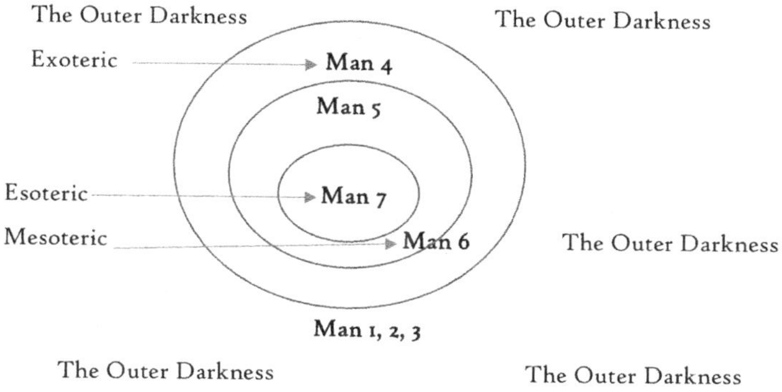

If an individual can become Man Number 5 he can still lose everything he has gained, but if he brings love into his life, the celestial functions become activated and a whole new world opens for him. The following diagram depicting the levels of man represents the realm of existence for each subsequent level of development.

The Outer Darkness is the realm of mechanicalness and psychospiritual sleep; this is the dwelling place of Man Number 1, 2 and 3. We are all born into this realm which is characterised by lack of understanding, emotional and psychological conflict, pain and unhappiness, general dissent, and war. Man Number 1 carries out physical tasks such as building roads, bridges, cities, and the like. Man Number 2 is artistic and perhaps composes music or paints. Man Number 3 is intellectual. He thinks, writes books and maybe carries out scientific work. These efforts though noble and worthy are nonetheless blind to the world of spiritual reality.

Man Number 4 is a little different because somehow, he has been stirred by the horror of the Outer Darkness. He has also acquired knowledge from a special source, resulting in the spiritual parts of his being becoming cleaved and unified. He lives in the Exoteric Circle. Man Number 4 receives influx, doctrines, and teachings from a higher level in an external way. What he receives he perceives truthfully and accurately, and he may eventually begin to discern that in the external method there is something hidden away. As a result, everything in him may change and undergo transformation and he may become Man Number 5.

Man Number 5 begins to think, breathe, and see in correspondences. If at this point in his development he keeps working on himself, he may come into the qualities and states of Man Number 6, but here he is on the border between Man Number 5 and Man Number 7. Although Man Number 6 has the perception of higher states he still has internal darkness which prevents him from progressing to the final stage of becoming Man Number 7, harmonised man.

Esoteric work insists that knowledge and being must keep pace with each other. When we die and pass into the spiritual universe our external knowledge disappears; it is our being which grants us entry into a particular spiritual society or community. What we are is our *being* and it is important to understand the distinction that knowledge is the property of personality, but understanding is a function of being. When we grasp the principles and understand the Scale of Gradation as a framework for work-on-oneself, we learn that the application of both the psychological and cosmological lines of development enable us, critically, to move from the lowest level of Man to the highest. And the direction in which we move determines our spiritual destiny, as it does the spiritual community to which we will ultimately belong.

The Work tells us that there are three principles we should apply to ourselves:

1. We should always be unconditionally sincere with ourselves. If we are observing ourselves, we must honestly observe everything that takes place inside.

2. We should never identify with any state that is observed. We must intentionally place ourselves to one side and observe our negative states, without self-condemnation or self-criticism.
3. Our observations must be couched in words that we accurately understand. We cannot be too general about this and must know the meaning of the words and terms that we use.

These three principles lead us into the three indissoluble rules of work-on-oneself; Self-Observation, Non-Identification and Self-Remembering.

To begin the process of self-observation it is necessary to set up within oneself a kind of observing platform or observation point from which to see our reactions and states in action. All we do is observe or watch whatever is going on. It may be thoughts, feelings, or reactions to other people or events. We don't try to change what is taking place; with focused attention we simply watch what is going on. It is important to be on guard for inner criticism, which is another negative reaction that works against us if we allow ourselves to identify with it. We are not to condemn ourselves regardless of what we see, and if we find that we are then we simply observe that tendency also.

The intention of self-observation is to get to know ourselves as objectively as possible. We don't think about what we are seeing or what we are feeling, nor do we criticise ourselves when we encounter something we don't like. We may register a thought or a feeling, but again we *see* the thought and we *see* the feeling and we don't put our sense of self into it, and the way we do this is by not identifying with it. It's quite dispassionate and somewhat like being in an audience watching yourself acting in a movie. You watch the character taking part in a series of scenes and gradually you begin to see that although each scene may have several characters playing various roles, the character you are watching always behaves the same. You begin to see that something is missing, although what that is, is at first hard to define. There seems to be a sense or feeling of lifelessness about the character, but even that is hard to pinpoint. It appears automated, like a machine which repetitively performs the same movements or actions, and because of your observation, you can predict how the character will behave in each scene.

There's another analogy which comes to mind. Consider the actions of a human hunter stalking prey. They follow tracks, they watch and study how the prey behaves at different times of the day and night, and they evaluate and determine its weaknesses and strengths. They watch how it obtains food and water, on which occasions and in what manner. And in time they know exactly how their prey will behave in accordance with its surroundings and environment, whatever they might be, and at the opportune moment the hunter will know when to strike. This is what we must do; watch with focused attention. Self-observation enables us to become the hunter. Our reactions and identifications become what we hunt. In effect, we are the character in the movie as we are the movie-goer in the audience, and we are as much the prey as we are the hunter who stalks it.

If we carry this out honestly, sincerely and without judgement, criticism, or any self-condemnation, we begin to see certain things about ourselves which simply can't be denied. We begin to see that we are not one 'I' but many, and it seems as though there are a host of others who live within us, each with their likes and dislikes, each with their own needs, wants, yearnings and so on. "Each minute, each moment, man is saying or thinking, 'I.' And each time his 'I' is different. Just now it was a thought, now it is a desire, now a sensation, now another thought, and so on, endlessly. *Man is a plurality*. Man's name is legion" (Gurdjieff, in Ouspensky, 1949, p. 59). In addition to Gurdjieff's perspective of plurality, perhaps we can also see ourselves within the context of being a multiplicity, and that one of the most effective means of knowing our multiplicity is by the action of intentional and focussed self-observation.

The act of observing is only an introduction, then comes the hard work of breaking up identifications and acting on negative habits. The Divine didn't create us to be mechanical, and as self-developing organisms it is our cosmic responsibility, and personal duty, to alter our positioning on the Scale of Gradation. There is something else that needs to be said about self-observation. Self-observation is just that and no more. We simply observe or watch ourselves in a gentle, quiet, yet intentionally directed manner. We don't try to analyse what we're seeing; we don't try to trace it back to our childhood or to a particular event that may have occurred at some point in our life. We must be careful not to use the Intellectual Centre in these situations because if we do work-on-oneself ceases at this point. If we start to analyse or question the origin of the state, then we have become trapped by it without even realising it.

This can be a difficult concept for the Western mind to embrace because it's quite different from much of what modern psychology, and counselling techniques, often encourage us to do. But this is what the Work requires of us, that we think about ourselves in a new way, a way so completely foreign to how we've been taught, that doing so quite literally reconfigures neural pathways and networks in the brain. When I entered the Work, I was Man Number 3, Intellectual Man, so coming to grips with this alternate way of understanding myself was extraordinarily difficult. I was always analysing myself and others, always telling myself that I behaved in a certain way because of whatever it was that had happened to me or because of what I had been exposed to as a child and it was only by repeatedly catching myself doing this that I was able to step back from it and observe the tendency. This self-analytical tendency had become such a habit and was so automatic that I despaired ever being free of it. But in time I did become free of it because through the application of self-observation and non-identification the tendency simply died away, or in Work ideology, became *quiescent.*

Returning to self-observation, if we find that we're running out of observing energy we may sink into the state and be overwhelmed by it. At this point it's necessary to concentrate deeply, wholly and solely on the Divine nature. We see that the Divine is the only reality, that the negativity has been created by us and that it is present within us. We try to feel, emotionally, that we are connected to the Divine. Higher Life will respond, and we will feel that response from the Divine as an impression from above, and when we do the negative state falls away instantly. We need release from these destructive states, and we must trust that the Divine will come to our aid. But we must ask for that aid; a door won't open unless it's knocked on first, and it is up to us to initiate that outreach.

Self-cleansing of the psyche is an important process in the Work. The only way to implement this process is to get to know those things we need to be cleansed of while observing them in action; this practical aspect of work-on-oneself is paramount. We cannot change these states, however by continually observing them we undermine them. If we can keep observing ourselves something from a higher source will act on them and in time they will fall away because they are placed in a state of quiescence. We must make these efforts, and we must be gentle, slow and persistent; change of being takes time and patience.

I grew up in an angry household. My father was suffocatingly controlling and as quick and loud to temper as he was to physically discipline and punish. I was aware from a young age of my frustration at having to be dependent for my wellbeing on both my parents and of being physically trapped in an environment I could not escape from for many years, two factors which contributed toward great unhappiness. When I became a parent myself, while I avoided the use of physical punishment as a means of disciplining my own child, inwardly I had internalised my father's anger. Fed by my own frustrations at life, the universe, and everything else that I could think of, I grew into a teenager and then an adult who was critical and intolerant of others, short-tempered and always angry. It was an awful state of being.

When I entered the Work I learnt that I had the ability to determine how I wanted to live my life, *now*. I needed to stop blaming other people and other things, and I needed to stop harking back to the past. Importantly, I needed to accept responsibility for how I wanted to live in the world. I had to abandon blame toward others and accept responsibility for my own internal reactions and thought patterns, and I did exactly that. I also had to change how I thought about myself, something which required a significant psychological shift, but when I applied this principle to myself, I discovered a different way of being in the world.

> Quite often in life we become negative and depressed. By sheer mechanical habit we tend to blame external events and other people for our negative states. We do not look at the situation clearly and automatically blame other people for our internal states. Our internal states should be under our control; if they become unruly, disorderly and generate unhappiness then it is clear that we are at fault. We are not obliged to allow negative states to rule us when things go wrong. If we are negative it is always our fault; other people are not to blame for our bad feelings. When this great truth is pondered, understood, and brought into practical

> application it can lead to a reversal inside ourselves. We see everything the other way around. (Groves, 1997b, p. 19).

Identification is one of the chief weaknesses of all human beings, and it's something that we do all the time. Whenever we say 'I' we fall asleep, and it's so automatic that we do it without even realising. This leads us to the process of non-identification, which like self-observation, is quite straight-forward. For example, suppose a particular event has triggered a negative emotion such as jealousy or resentment. Self-observation enables us to see the reaction, that's the first step. By actively observing the state we automatically move to non-identification. That's the second step. It's important to understand that as soon as we attach 'I' to any state we become that state, but by intentionally separating ourselves from it we change everything.

We do this by making a conscious effort to say to ourselves "I observe a state of [emotion]", which is vastly different to "I am [emotion]." We don't put our sense of self into it, instead we view it from an internal observation platform. It is a mental shift that though subtle, produces immediate results. Another way to approach this is to say to ourselves that 'it' is reacting, 'it' being the personality and its multitude of 'I' states.

We often forget that we have a choice. We can choose to become the state, or we can choose not to become it, it really is that simple. Although I like to consider life in the sense that it isn't metaphorically black and white, in this instance it is. We can choose to identify with the state or not. It's up to us to exercise that choice, and it's up to us to make that intentional effort. There is something else to be considered here. We need to be careful not to identify with the act of observing, and this is where the act of self-remembering, the third indissoluble principle of work-on-oneself assumes its role in the psychodrama of our ongoing transformation.

The pendulum of our emotions and their reactions swings within our being from low levels to high and from high levels to low. If we identify with the swing of the pendulum, we live our lives in accordance with these things and remain at their mercy. We attain a state of equilibrium when we attempt to be in charge of the very things that cause the pendulum to swing, by observing and not identifying with them. In a sense, equilibrium is the platform or level of being which allows a person to consciously act rather than mechanically react to the events of life.

There are some states with which it is necessary and right to identify. States such as love for God, awe at the ineffable, wonder at the miracle and diversity of life in all its forms and manifestations. Humility, when we see our nothingness in the face of the balanced harmony of the living universe, but above all, love, love for God, for the Work, for the Word and for the Neighbour. These are essence states which are far removed from the emotional reactions that have their origin in the personality.

In time, and because of the application of non-identification, the intensity of repeating psychological states and negative emotional reactions lessens. This is because we're not keeping them alive by continually putting our sense of self into them. Gradually we notice that we're not troubled by them to the degree we once were because they have become quiescent. Every so often they may rise again to the surface, and as before we observe and don't identify with them. We must be stubborn about this in the positive sense of the adjective's meaning. This is what is termed 'Metanoia'. A transliteration of the Greek terminology, μετάνοια, means 'change of mind, change of state'. Metanoia is the process of going beyond the ordinary mechanical mind. It is an ongoing and lifelong process and involves two types of thinking: psychological thinking and cosmological thinking.

Psychological thinking incorporates the Work and the practice of observing ourselves. We must ask ourselves; am I sincere? What did I discover? We think about outer considering and inner considering. We think about the nature of impressions, the need for rebirth or transformation, and our spiritual destiny. Cosmological thinking involves thinking about cosmic scales such as the Ray of Creation, the Side Octave of the Sun, cosmic energies which we can attune ourselves to, and influx from the Divine. The process of Metanoia changes the mind because it (the mind) is used differently. The Work teaches how to think in a new way and how to use the mind in a new way. Self-observation and non-identification are important components of this, and when properly utilised enable us to extend ourselves beyond the mechanical mind.

Occasionally there is an influx from the Divine into suitable minds of a new idea or a new way of understanding the world, such as for example the notion of neuroplasticity. Up until relatively recently, an accepted scientific view in the West was that the adult brain was immutable in two respects; that new neurones (a process termed neurogenesis) were not generated beyond a certain age and that functions of the various structures of the brain were determinate that is, unchanging. However, in the early 19th-20th Century Psychologist William James introduced the term 'plasticity', which was a forerunner for both the term and science of neuroplasticity, in his 2-volume work *Principles of Psychology*:

> Plasticity, then, in the wide sense of the word, means the possession of a structure weak enough to yield to an influence, but strong enough not to yield all at once. Each relatively stable phase of equilibrium in such a structure is marked by what we may call a new set

> of habits. Organic matter, especially nervous tissue, seems endowed with a very extraordinary degree of plasticity of this sort; so that we may without hesitation lay down as our first proposition the following, that *the phenomena of habit in living beings are due to the plasticity* of the organic materials of which their bodies are composed.* (Vol 1, Chapter 4, p. 105, 1950).
>
> [*In the sense above explained, which applies to inner structure as well as to outer form.]

Scientific research has now revealed that the brain is a highly adaptable organ and that its functionality is not fixed. For example, the science and study of neurogenesis (the development and growth of new neurones or brain cells) has revealed that this process can occur in areas of the brain from the pre-natal stage up to adulthood and potentially across the lifespan. In adulthood, new neurones are created from stress-responses to, for example, training or exercise, or to activities which generally stimulate the brain in a positive way either physically, emotionally, or psychologically. This suggests that the brain can remodel or adapt itself to accommodate changes or new experiences which occur in an individual's environment.

Like neurogenesis, the science of neuroplasticity has in turn given birth to the emerging field of the study of spiritual neuroscience. Spiritual neuroscience examines correlations between neural phenomena and the subjective or lived experience of non-material phenomena deemed spiritual, and neurotheology, the study of the relationship between the mind and religion (see for example *Principles of Neurotheology*, Andrew B. Newberg, Ashgate Publishing, 2010). While these important emerging trends reveal a cause-and-effect principle, it is also important to remember that the Work consists of *spiritual* ideas which teach us how to *spiritually* grow and evolve.

While the Sciences have been of enormous benefit to mankind in profound ways, as a discipline it has also had its proponents who have used a paradigm of empiricism to define 'reality' and to decry the spiritual origins of reality and the universe.

The mystic on the other hand, might propose perhaps that when encountering non-material or spiritual phenomena, that it is the *feelings* generated by such experiences which are the reality because they confirm it. This occurs because in that moment a truth is not only experienced, seen and known in ways which cannot be verbalised, but because the spiritual and cosmic worldview of the individual changes in response to the experience. The material world is the end-effect of an immensely creative process which is spiritual and Divine, and we must be cautious in believing that only science can yield reliable knowledge and truth about the nature of reality.

> The hidden things of spiritual writings are the esoteric, arcane or celestial-spiritual contents and it requires a certain degree of inner awakening to see and appreciate them. Such spiritual truths are timeless and never go out of date. Truths recorded 5,000 years ago are still fresh, sparkling and alive today because they come from the timeless realm of the eternal and cannot fade, or lose their potency. Therefore God-given spiritual truths written down millennia ago are specifically for us, the readers of this time. If we read with purpose and are receptive, God directly instructs from the internal things within the written word. (*Author's lecture notes, 15 March 1995*).

Self-remembering is the third insoluble principle of work-on-oneself. What do we remember? How is it done? According to a Sufi teaching, remembering is "an active recollection, a gathering of oneself together and making oneself conscious of eternal verities"

(*Author's lecture notes, 17 December 1997*). 'Gathering oneself together' means knowing how to integrate ourselves and knowing what elements and principles must be incorporated into that pattern of integration. 'Eternal verities' include the reality and presence of God, the emanations which perpetually proceed from God, the laws and powers of world-building and world-maintenance, and the activities and influences of communities of evolved souls. It is the process of looking out to see the influence of God in all things while also looking in to see the presence of the Divine in ourselves, and both are carried out simultaneously.

Self-remembering is a technical term which is repeatedly found throughout esoteric writing and teaching, and although self-observation, non-identification and self-remembering are interdependent with one another, it is the activity of self-remembering around which work-on-oneself revolves. "The whole of the Work is based on Self-Remembering and we cannot awaken to greater life unless it is carried out" (Groves, 1997d, p. 17).

In an initial draft of the manuscript, and when writing about self-remembering I prefaced the first paragraph with the following words, "I must write with honesty that for some reason I have had difficulty understanding exactly what is meant by self-remembering, first I think because of the term itself, and second because there appear to be different degrees or types of self-remembering." The paragraphs following that statement clearly evidenced my well intentioned though grasping understanding of the term, as well as my ignorance of the implications of self-remembering. Re-reading what I had written eighteen years ago and comparing it to my understanding now, I can see how limited my comprehension and understanding of the process was at that time.

In an attempt to find my way through the maze of misunderstanding, I returned to my original lecture notes and read the following:

> Which 'I's' are involved in self-remembering? This is confusing because the outer mind is thinking about it. Self-remembering is remembering that there is a bigger part of ourselves that we must reach. This takes years. We cannot suddenly move from the outer mind to the inner mind, we must realise this. We have an enormous amount of inertia which we must struggle against. (*Author's lecture notes, 15 October 1997*).

This is helpful because we see that various factors are involved in the process of self-remembering. Continuing my re-reading, I found myself pondering on the following references which had been made by Dr Groves during the lecture which referred to Dr Maurice Nicoll, Emanuel Swedenborg and P. D. Ouspensky's experiences of self-remembering:

> All real self-remembering is simply forgetting yourself, your ordinary self ... No one can remember himself unless he feels that there is something higher than himself. Unless he feels this, his self-remembering will always lead him, into False Personality. (*Author's lecture notes 15 October 1997 referring to Nicoll*).

> As often as I repeated the prayer of our Lord, morning and evening, almost as often was I taken up in various ways into the more interior sphere, and indeed with such a perceptible change that nothing could be more perceptible. Explanations of the more interior things of the prayer were then insinuated with very great variety, but when the prayer finished I was remitted into my ordinary sphere. (*Author's lecture notes, 15 October 1997 referring to Swedenborg*).

> Ouspensky also wrote about self-remembering. He describes it as a process which involves looking out at the world and in toward the self simultaneously. Everybody is different, and what works for one may not be suitable for another. (*Author's lecture notes, 15 October 1997 referring to Groves*).

It is difficult to mentally remember to bring together and unite the components of self-remembering such that they form a coherent whole in the understanding and doing so takes time. Both Ouspensky and Groves shine light on two important considerations; that the elevated states of self-remembering are temporary (but if we continue to carry this out we build up a new level within ourselves), and the uniqueness of technique (what works for one may not for another). It seems that we have to outwit ourselves when carrying out self-remembering activities and exercises, otherwise our efforts become mechanical. For example, on one occasion we may see ourselves standing in a doorway between two rooms while feeling that we are standing in the two rooms simultaneously, or when in one state we may remember its opposite; when happy remember unhappy, when tired remember boundless energy.

Self-remembering is most potent when it becomes emotional, furthermore, it is only occurring when we *feel* the reality of the process. Say for example we are encountering an impression. It may be a flower or a pleasant scene, words on the page of a book, another person, an ant crawling on a twig, a microbe under a microscope; it can be anything. Because we encounter these impressions with the energy of awareness, the experience is one-dimensional. This alters dramatically however if from another vantage point within ourselves we are able to watch ourselves observing the impression. When this occurs our sense of self shifts and a dramatic and instantaneous experience of being the observer and being the observer observing the observer takes place.

It is difficult to grasp and understand the concept that 'another me can observe me and I am both simultaneously', but when the experience occurs the feeling that one is simultaneously present in two realities is undeniable, and it is undeniable because that is exactly what is taking place. Sometimes this experience occurs when dreaming, especially in those dreams where the dreamer feels they are awake in the dream (as they would be during the day and going about their duties) yet knows in their mind that they are asleep; a contradiction to be sure. There is a definite sense of being in yourself dreaming and being in your body, coupled with the sense of being fully awake in the dream as you would be if you actually were. And while this is transpiring, the dreamer is aware in their mind that they're dreaming, which serves as a bridge connecting the two states of being. It is dynamic, highly fluid, and difficult to express in words, but when you experience it, you understand it.

We must always remember that feelings are the experience of the reality. The observer observing the observer is an 'I' which stands at a higher level than those 'I's' who observe or see the initial impression; this is Real 'I'. These levels have a particular energy associated with them and when the experience is over, a residue of the energy from both levels, which are quite different to one another, remains.

Many years ago, I had been pondering on self-remembering and was asking myself such questions as, What is self-remembering? What does it feel like? How does the experience occur? How do you look in and out simultaneously? As usual I couldn't see the obvious until one day quite unexpectedly, I experienced a higher essence state of self-remembering.

I was walking down a busy street on my way to buy some lunch.

I was aware of the differing tones of male and female voices and the sounds of traffic; squeal of brakes, engines revving, horns tooting. Chirping and twittering from birds flying overhead and resting on tree branches filtered down to my ears, mingling and adding to the cacophony of sound which bombarded my senses. I noticed the movements of my body, the swing of my arms, my breath in and out, my heart beating in my chest, the impact of my feet with the pavement, the sound of my shoes.

As it was lunchtime the cafes were full. The clink of glasses vied with the odour of cigarette and cigar smoke. I walked past well-dressed men and women with sleek hairstyles and bright lips and glanced briefly up at the cloudless sky which formed a brilliant blue canopy over the entire scene. I thought idly to myself that the blue was the same as that in the spiritual world. There was a brisk breeze which when it touched my skin sent goosebumps rippling up my arms from wrist to shoulder. I perceived that I was in a heightened state of awareness, both of my surroundings and myself, when suddenly a most extraordinary thing occurred.

I felt myself being drawn up and out of myself so that I was within myself yet up and away from myself as well. Then I could see myself seeing myself. This was astonishing beyond belief because I could actually see myself seeing myself. But it wasn't just seeing myself, it was *feeling* myself seeing myself. At that moment I felt that I was in another place and that I was experiencing another kind of reality. I understood that what was taking place was an experiential event of another kind of existence altogether. And then just as suddenly as it began it was over and I was back in myself looking out at the world in my usual manner.

The very next day the same thing happened, and again I was drawn up and out of myself to find that I was seeing myself see myself. As before it wasn't just experiencing this, it was *feeling* it which was the most astonishingly profound aspect of the event. I could feel that what was occurring was a living process within my being. I experienced and felt two different realities simultaneously because I *was* in two different realities, two different worlds simultaneously, and again I felt that I was experiencing a different kind of existence.

There was an intensity about this, an exquisite bliss which made it almost impossible to bear, and then just as suddenly as I was drawn up and out of myself I was let back down into myself to find myself looking out at the world in my usual manner.

Self-remembering is the experience of consciously knowing *our whole being* and seeing that it depends utterly on higher things for its existence. "Self-Remembering is the enlargement of the field of consciousness so that both the outside and the inside worlds are grasped together in the unity of experience, and it is then that man obtains an inside view of himself and everything around him" (Groves, 1996c, pp. 1-2). Higher Life continually seeks to educate and inform us, and in constantly striving to bring us to our greatest happiness will utilise temporal space and time, as demonstrated in the two higher essence experiences previously recounted, in unique and creative ways to do so. And thank goodness it does.

Gurdjieff referred to the fact that human beings cannot 'do', rather they have things done to them. "A man who does not possess his own world can never do anything from his own initiative; all his actions "are done" in him" (Gurdjieff, 1991, p. 173). The higher essence experiences that occurred to me were something over which I had no control. Both instances were spontaneous and occurred randomly and the truth is I didn't 'do' anything to make them happen; in the sense of Gurdjieff's statement, they were done to me.

There is a corollary to this, as there often is in the Work, and it's somewhat contradictory, as they often are. While I didn't 'do' in the sense that I personally harnessed and directed an internal power to make this higher essence state of self-remembering take place, what I did do was initiate the process by first turning my attention toward considering it, and what I can lay claim to is the effort required *to want to know and understand what self-remembering is,* after which Higher Life responded. Perhaps how others come to their understandings of self-remembering might be different to mine. The Divine is both resourceful and creative and it seems that in this instance it took an event like that described to get through the barrier of my outer self.

A loving relationship between the Divine and ourselves and our acknowledgement of our dependency on the Divine, within the context of, 'of myself I can do nothing but with God all things are possible', is crucial to work-on-oneself. Framing and contextualising this relationship are several elements which the experience I had reveals:

1. We must have an aim.
2. We must make efforts to realise the aim.
3. Higher Life will respond to those efforts.

We need to understand that without intentional effort nothing is achieved, and we must play our role in that effort with sincerity. I have lost count of the number of events that have occurred in my life where I have felt that Higher Life quite deliberately responded to my outreach toward it to communicate, educate and inform me about myself and the reality of the multidimensional universe in which we live. I recall a conversation between Dr Groves and myself during which he told me that an enormous amount of life experience was being condensed within me in a short space of time and this was also a reason for the diversity of spiritual experiences which had populated my life.

In 1987 when I was in my early twenties, I travelled for nine months in Great Britain, Europe, Greece, and Turkey. In July I spent two weeks in Turkey, my intention being, amongst other things, to visit the tomb of Jalal ad-Din Muhammad Rumi in Konya. One cannot ever visit such an exotic locale without tales to tell and the following account, which is one of many that I accumulated during my journey abroad, reflects that diversity to which Dr Groves was referring.

It was late afternoon when we finally arrived at the tomb and mosque which housed the remains of Rumi. As I walked toward the entrance dozens of impressions struck me at once; the dust swirling around my feet, my long skirt wrapping itself around my legs, the heat of the ever-present sun on my back, the sound of a hundred voices, women swathed in black, their eyes the only indication that beneath their voluminous robes there actually existed a human being, and the men, everywhere, either huddled in groups or sitting, squatting, standing, staring. I didn't know what to expect. My heart began to thump forcefully in my chest as I left my shoes at the door, wondering fleetingly how on earth I would ever find them again amongst the hundreds of pairs that littered the entrance.

And then I was out of the glaring heat and standing in a line with people from what seemed to be every country on the globe. French, German, American, Italian, Turkish and English voices all converged into a single sound which became hushed as we entered the tomb. And then, my God the stench, the stench was unbelievable. The stench of thousands of unwashed feet assailed my nostrils with such ferocity that I thought I was going to vomit. But nobody else seemed to notice, or if they did, had the good manners not to cough or hold a hanky up to their nose, as I was doing. I managed to calm my stomach and continued inside.

The walls were painted with some of the most beautiful gold lettering I had ever seen, with great swirling floral patterns and designs brilliantly coloured in blue, turquoise and red. Carpets covered the floor and from the ceiling hung great lamps which bathed the room with a soft warm glow. There were people everywhere and with great dismay I saw the sarcophagus, thronged with people 4-deep forming what seemed to be an impenetrable barrier; I would never get close to it. Everyone towered over me, how could I get through them? What was I to do? I looked toward the entrance seeing more people enter by the minute, everyone surged forward, but me. But then, as I stood there thinking those very thoughts the people seemed to melt away and unbelievably, I found myself standing in front of the sarcophagus, alone, with not another person nearby.

Suddenly the voices became a distant drone, and I felt a latihan state come over me and then I felt something descend from above, an energy of some sort. I felt it pass through the top of my head and melt into my body, felt it become a part of me, felt it make its home in me. I felt Rumi's presence as a silent witness to this act and bowed my head and then just as suddenly as it commenced it was over and the voices became loud and intrusive and that overwhelming stench again, and everything rushed rudely back into my awareness bringing me with a jolt back to material reality. Suddenly I was surrounded by people. I was aware of them brushing against me, of their exclamations of wonder, of women stopping to pray, of black robes entwining with my own garments, of the smell of perfume and acrid sweat and the clink of jewellery on slender arms. But most of all I was aware of hearts reaching out to God, of souls yearning for the company of other souls, and of the overwhelming mystery that is, and what will always be, the journey to God.

I don't really know what happened. What I do know is that I implicitly trust in that which orchestrated this event, because something did, and it wasn't me. I know the foundation for this trust was laid in my childhood and in the love which I continually felt seeping into me. I have no control over random events like this, I never have, and I know I never will. I don't recite incantations, I don't conduct little rituals, I have never tried to force the hand of Higher Life and I have never bargained with Higher Life. Because these events occur it doesn't mean that I'm special in any way; I don't have an egotistic presumption that I'm a 'chosen being', it's just the way life is. Irrespective of the presence of these experiences in my life, I know what I am, and I suffer no less and struggle no less than any other person who works on themselves.

I recall on another occasion, when I had spoken with Dr Groves about the intensity of feelings that accompanied such experiences, he told me that my "exquisite spiritual sensitivity" was the price I paid for them. "Nothing is for free here", he had said gently and with great compassion. "There is a price to be paid for everything and your sensitivity is the price you pay" *(Dr Philip W. Groves, personal communication, 7 June 1997)*. Apparently, this sensitivity has also contributed toward long years of having to battle constant bouts of ill-health, unusual illnesses, and physical pain.

I'm reminded of an educational experience which occurred on a crowded train when I was travelling home one afternoon from work. I had been reading one of Emanuel Swedenborg's books and for a few days had been pondering on the fact that infernal spirits can insinuate themselves into a person such that they manifest physical sickness and ill-health. It was standing room only on the train and as I was observing my body sway in time with the rhythm of the carriage, I suddenly saw a dark shadowy shape, unseen by others, moving intently toward me. Here we go, I thought to myself.

When it entered my body, I doubled over in pain as an excruciating twisting sensation gripped my lower abdomen taking my breath away. Although I was experiencing great discomfort, I knew intuitively what was taking place was an example of exactly what I had been reading about. When I arrived home the unexpected visitor was soon dispelled by a latihan, which took care of the spiritual infestation, followed by copious quantities of slippery elm powder, which soothed impacted and inflamed tissues. I have always been a willing participant in the spiritual events that have occurred in my life, even those which may have been a little jarring because I understood them for what they were; educational experiences. So to have I always felt that something was being *done* to me and that feeling has never changed. Meaning that I did not initiate them or plan them, they just randomly occurred. It is diagnostic to explore the feelings that accompany such events, because it's the feelings that confirm the reality. And even if confronted by public or private ridicule, such as I have, we must never lack the courage of our convictions of what we *know*.

In November 2008 I was an invited speaker at a state Palliative Care conference. I was in the early throes of completing my PhD and my presentation, *Post-death contact, continuing bonds and grief* discussed after-death contact experiences between the bereaved and those close to them who died. I was arguing that these experiences, rather than being pathological are in fact a key element of the continuing bonds model of grieving. This model posits that in order to come to terms with a death and the grief that follows, the bereaved don't need to let go of or detach themselves from the deceased, rather they maintain a bond with them in some way shape or form. I can't quite describe the looks on the faces in the audience, although I do remember the face of one woman in particular; she was aghast.

Nobody asked any questions, which was not unusual, however afterward one person did approach me. He was taken aback by several aspects of my presentation, including the subject matter and the presenter, me. We spoke for about 30 minutes but all I recall from the conversation are two sentences. His: "You look so normal." Mine: "I am normal." The inference for him was that the dead are not normal, nor is the afterlife, nor is someone who believes in those things much less speaks publicly about them. For me of course all that is 'normal'.

The process of completing my PhD, *Ways of Being: The alchemy of bereavement and communiqué* (Knight, 2013), heralded my speaking publicly for the first time about my experiences of the afterlife and multidimensional reality, but it has come at a cost; ridicule, academic jealousy, avoidance by others and lack of post-doctoral opportunities. The road has not been an easy one, but I wouldn't change any aspect of it because I know what I'm a part of, and I know where my spiritual life is leading. My existence here as an embodied being, as a planetary dweller, is effectively a blink of time, no more and no less. It is an important waystation for entry into the spiritual universe, into eternity, and that is all.

Work-on-oneself can be conceptualised as an ascending octave, and can be placed on the Scale of Octaves as a process of self-development in the following manner:

Do Union of outer self with inner self – Integration

Point of Blockage

Si Awakening of fine emotions
La Cleansing of the outer self
Sol Efforts to deepen the Work
Fa Metanoia
Mi Learning to make initial efforts
Re Grasping the idea of self-development
Do Introduction to Work ideas

Our introduction to the Work sounds the note **Do**. When we grasp the idea of self-development, we sound the second note **Re**. **Mi** is sounded when we undergo the process of learning to make initial efforts such as observing ourselves and not identifying with our reactions. Then we come to the first point of blockage or gap in the octave. Gaps in the octave generally represent psychological difficulties. This is an active and dynamic point because it involves rethinking our life in terms of Work ideas. This is like a revolution in the mind.

The process of Metanoia, going beyond the ordinary mechanical mind, sounds the next note, **Fa**. Our efforts to deepen the Work, to understand it, to digest it, to make it organic sound the note **Sol**. Cleansing of the outer self through self-observation and non-identification sounds the note **La**. **Si**, the next note in the octave is sounded with the awakening of fine emotions such as love for the neighbour, charity of spirit and so on.

We then face the second point of blockage or gap in the octave. This is the point at which the energy of love is allowed to flow into a person where it is cherished. When we sound the final note in the octave **Do**, we achieve integration. The outer self and the inner self have become unified. We have become a complete being with the result that our pathway to Higher Life opens up.

Work-on-oneself is an ascension toward Higher Life. The notes of the octave are stages in the spiritual development of a person who is working on themselves. The scale itself is a representation of the Divine idea of Man as a self-developing organism. Human beings have not just come into being as the result of biological evolution. There is a spiritual pattern which lays the framework for the nucleated soul of the essence which is brought forth by Divine will. There is another history of mankind that has been overlooked and ignored by the scholars and scientists of the day, and it must be plainly stated that man's birth on this planet is by design, not accident.

I recall reading somewhere that Emanuel Swedenborg contextualised the human race as being the seedbed of heaven. From a spiritual perspective we are born for a specific purpose and reason, and we carry within ourselves the possibilities and capabilities for spiritual growth and development. We need to know how to maximise this potential and to similarly understand that we are self-evolving organisms which need special knowledge from a special source in order to do so. If we are going to start at all, we have to start at the bottom rung of the ladder, at the first note of the octave.

> One of the great weaknesses of the human mind is its mechanical tendency to assume that it understands what it hears for the first time. Sometimes the mind can understand something it hears only after the hundredth time! Students repeatedly fail to grasp important details of what the Work teaches because of the foregoing assumption. This mechanical tendency must be countered by repeatedly going over earlier Work material with a 'fine tooth comb' with the purpose of examining, exploring and expanding various details until they become expositions of practical Work principles. Only in-depth knowledge of Work teachings can help students to acquire an in-depth understanding of their own being. (Groves, 1997e, p. 18).

As the quote by Groves suggests, if we are going to help ourselves, we need to understand the necessity of revisiting Work basics time and again throughout our lives. And the way we do that is to remember the triad of Self-Observation, Non-Identification and Self-Remembering. When we bring these three activities together, they will work within us as a unity.

6

FAITH

Is there some kind of knowing that can transform our being to the point – dare we imagine – of bringing forth a life that does not die when the body dies? Such knowing as this is inseparable from the action of faith – considered not simply as a set of emotionally charged beliefs, but as a movement within the human psyche that generates a magnetic current flowing between our individual human life and the source of human life itself; and that deposits into our human life the spiritualized matter of what is called the "new Adam"

... or in other words, to answer with the actions of love rather than with brittle promises of future virtue. Such transformational knowing actually has little or nothing to do with what we ordinarily call thought. It has to do with energy, the energy of consciousness.

Jacob Needleman, Foreword, *The Gospel of Thomas*, vii-viii, 2005.

There are three functions of the essence, Hope, Charity and Faith and one of those functions gives its name to the title of this chapter, Faith. How do we understand faith from an esoteric perspective? The Work tells us that faith is a function in the psyche which begins its life in potential form. With the addition of love and trust that potential is maximised so that in time it becomes active and dynamic. The following equation demonstrates the principle:

Faith as a *potential*, + Love + Trust = Faith as a *function in the psyche*

In the context of the Work, faith is a technical term, and in the alchemical process of work-on-oneself it is love and trust which is transmuted into faith. Faith is not merely blind belief, but a psychospiritual process which leads to a state of being. Faith is also an octave of inner development which contributes to our ascension toward Higher Life. Being part of the process of soul evolution, faith combines the process of growing in love toward God and surrender to God which in turn walks, metaphorically, hand-in-hand with trust.

This is something that takes time, however we can be encouraged in our efforts by knowing that faith, as a function in the psyche, when active, changes our being. Many would be familiar with the words of Paul the Apostle, "Now faith is the substance of things being hoped, the evidence of things not being seen" (Hebrews 11:1). Pondering on these words reveals the essence of faith which asks us to believe and trust in something which can't be seen, which is God. The key to deciphering the inner meaning is to understand that faith isn't about seeing God with our eyes, it's about feeling God with our being, and the pathway to that is through a loving relationship with the Divine.

We need to conjoin this understanding with a deliberate mental shift away from our initial confusion when we first encounter Paul's words because there is an important subtlety here which is at risk of being overlooked. We can't see God as we would see something material, yet that doesn't mean that God doesn't exist or is not present. We don't 'see' the oxygen we breathe either, yet it exists.

We don't 'see' microwaves cooking food, yet the food cooks. This is part of the *process* of faith, not to see yet believe. But more than that, it's not to see yet *know* and *feel* the reality of God, which is more profound than simply giving mental assent to the notion that God or a higher power exists.

While faith as a process incorporates trust and love, it also requires surrender. We trust in that which we cannot see, we love that which we cannot see, and we surrender to that which we cannot see. Faith as an activity requires that we abandon our habitual and mechanical way of interpreting and understanding the world and our habitual and mechanical way of interpreting and understanding faith. Faith as a function in the psyche is an attainable level of being and a meaningful way in which to live and participate in life, which in turn orientates us toward a heavenly direction. We see an example of this in the writings of Mohandas Karamchand Gandhi (1869-1948), otherwise known as Mahatma Gandhi, India's well known political social justice activist and advocate for non-violent civil resistance. A devout Hindu, Gandhi was able to draw from all religions one enduring truth: God as the source of all life. He saw that the world's religions were different paths which all led to the one undeniable reality:

> I do have a living faith in a living God. A living immovable faith is all that is required for reaching the full spiritual height attainable by human beings. God is not outside this earthly case of ours … we can feel Him if we but withdraw ourselves from the senses …
>
> It is faith that steers us through stormy seas, faith that moves mountains, and faith that jumps across the ocean. That faith is nothing but a living, wide-awake consciousness of God within. He who has achieved

> that faith wants nothing. Bodily diseased, he is spiritually healthy; physically poor, he rolls in spiritual riches. (Kamath, 2007, pp. 131-133).

I recall one time having a conversation with a Catholic priest about faith. I said to him matter of factly that he must find many occasions whereby his faith is tested. He readily agreed. Afterward I reflected on our conversation, particularly my assumption that the attainment of faith was predicated on it being tested. Something about my assumption didn't feel right, which is why I had turned my attention toward it. Suddenly I realised that if I had faith then that faith shouldn't need to be tested to determine whether or not I had it in the first place. I found myself confronted by what seemed a confusing contradiction. I believed that having my faith tested was integral to having faith, and that such a test was required for my faith to grow.

Faith isn't evidenced or proved by tests, nor does it 'grow' in this manner, and what I realised after my discussion with the priest was that what I was seeing was my own doubt and lack of trust being brought to the surface. Clearly faith wasn't yet an active function in my psyche. Faith is not mere mental assent and acknowledgement of the thought of God's presence, nor blind or mechanical belief in that presence; it's an active rejection of our external sensory perception of reality. Faith is an *outcome*, and it results from our ability to bring something together within ourselves which exists beyond the limitation of the senses and their perception of that reality.

An illustration of this can be found in the small work *De Miraculis*, written by Emanuel Swedenborg, who describes an encounter with spirits who were discussing miracles. Some of them who had no faith regarding life after death or of the internal sense of the Word wished to see a miracle because then they would be convinced, then they would believe. As Swedenborg points out however, the truth is otherwise:

> But the answer to them is that for those who have no faith, miracles can accomplish absolutely nothing; while for those who have faith they are unnecessary. The reason for this was stated, namely, that faith is not implanted by miracles, that it is merely strengthened to some extent ... The reason why plainly-apparent Divine miracles do not now take place has been revealed to me. It is because the inward things of faith, which we receive from the Lord, cannot be sown or implanted under compulsion, but only in freedom, thus not amid the terror and amazement induced by miracles. (Johnson, 1943/1947, pp. 5-7).

Faith isn't born in the maelstrom of life; it's a process that plays an important role in how we live spiritually and how we perceive and understand reality. It is something within us that strengthens over time, quietly, until we reach the moment where we know that over and above all things, God *is*, always *has been* and always *will be*, and it is this constant which sustains us. This was something I was very aware of during the breakdown of my second marriage, which extended over a bleak eighteen-month period. Although this period was characterised by painful emotional and psychological turmoil and confusion, as an event in my book of life so to speak, it provided me with the very material that I needed to work on myself, particularly my understanding of the function of faith as the following excerpt from my journal reveals:

> Over the past few days I have been reflecting upon the events of the past 18 months. The events have been tumultuous to the enth degree. With each successive

event I recall feeling that something was being stripped away from me until it seemed that everything I held dear was no more. I felt that I was being shown that all the things that I had thought constituted my life were in fact an illusion. That all the things that I believed held it together, giving it shape and definition and providing security and comfort, were in fact not the foundation for it at all.

I have come to see that throughout the months of what seemed to be endless turmoil, something was sustaining me. That throughout every second, a power greater than myself was responsible for me. That something other than what I believed had provided the structure and framework for my life, was in fact the very foundation for my ableness-to-be. I know this beyond a shadow of a doubt. I simply know it.

I see that I am nothing compared to the something that is everything. I see that I don't really exist at all, that my existence, as I believed it to be, was a fallacy, an illusion and that in a very real sense, I am not real. There is only God. How can one verbalise that which is known in the heart? One can only experience the meaning of it. I was self-centred, not God-centred. Dr Groves had spoken about this in class; I remember not understanding his words and wondering at their meaning. Now it has been shown to me through direct experience.

I feel that there is a benevolence that runs through all life. Like a great flowing stream it proceeds from the Divine. It is good and kind and loving and passes

> through all souls, irrespective of whether they are aware or unaware of it. There *is* an essential goodness in life which permeates and saturates all existence by its presence. Despite the spiritual degeneration of mankind this benevolence exists, weaving its way through the cosmos, touching the hearts and vivifying the souls of all created beings.
>
> It comes from a source that is great beyond measure, that no word can describe or sound emulate or mind possibly conceive. I am nothing in comparison to that something. There is something truly good that maintains the all; life as we humans know it, the universe in its entirety, and the essence of creation itself. *(Author's journal notes, 16 May 2001).*

When we work on ourselves, we do so in the midst of daily life, and it is life events, with all their joys and sorrows and twists and turns which provide the very material we need. From an external perspective, the dissolution of the marriage had a flow-on effect which not only impacted me in various ways but also my son, who was only a young child at the time. Interiorly however the landscape was vastly different, which is reflected in the journal entry. It was faith, as a conviction of the reality that I couldn't see, which sustained me during that time.

Unfortunately, a literal understanding of faith has contributed toward a common misconception held by scores of people that salvation is obtained through faith alone. Many people in positions of religious authority and power have promulgated this doctrine, which has done nothing for the inner life of those unfortunate enough to have come under its influence. The Work tells us very clearly that our spiritual destiny is determined, shaped and governed by the manner in which we have lived our life, and that self-cleansing is an essential and life-long activity. A literal understanding of faith is not enough to ensure salvation. Swedenborg cautions:

> There is a general belief today that all are saved who, at the close of life, have certain pious thoughts and from their heart speak about salvation and concerning the Lord, especially if they then confess that the Lord suffered for them. Such teachings may afford some consolation to those at the point of death, they may allay anxious thought about one's previous life, but the truth is quite otherwise. It is the previous life that determines a man's future happiness or unhappiness, and piety just before the hour of death does not take anything away from that life. (Johnson, 1943/1947, p. 17).

Because of the literal understanding of the doctrine of salvation through faith alone, countless numbers of disembodied people are streaming into (not to mention those who have already entered) the spiritual universe in an untransformed and incomplete state. This common perception of faith is inadequate and akin to spiritual laziness. *We* are responsible for making the effort to clean up our own psychospiritual mess, no one can or should do it for us, and to think that stating certain words or giving mental assent to them at the hour of our death will somehow wipe the slate clean, well that's simply ludicrous. Christ stressed the importance of self-cleansing using the Pharisees by way of example:

> Woe to you scribes and Pharisees, hypocrites! For you cleanse the outside of the cup and the dish, but within they are full of robbery and excess. Blind Pharisee; cleanse first the inside of the cup and of the dish that

> the outside of them may become clean. (Matthew 23:25-26; See also Luke 11:39).

This is sound esoteric instruction and practical religion and is a direct reference to work-on-oneself. We are the Pharisee as we are the cup and the dish, and we must first clean the inside of that cup and dish, which corresponds to self-cleansing. The instruction provided by Christ is direct and clear. Rather than our external behaviour mirroring piousness and religious or spiritual duty, we must cleanse our internals to bring about *change of being*.

When in conversation on another occasion with a group of spirits, Swedenborg relates an encounter in which he attempts to discuss the truths of faith. His questions elicit no response from the spirits he is speaking with, and he eventually realises that it is impossible for them to have any opinion or belief regarding faith. This is because when living as embodied beings they didn't believe in the existence of anything of a spiritual nature, including heaven, hell, or life after death. To this end, what constituted their life before death, that is their thoughts, opinions, and beliefs, remained as such after death. Swedenborg writes that because they only believed in what was natural before death, they subsequently believed only in what was natural after death "... and from this belief, a state of this kind awaits them after death" (Johnson, 1943/1947, pp. 9-11).

Swedenborg illustrates a fundamental principle of work-on-oneself. We must undergo transformation here and now because what we have made of ourselves in this life is what we take with us into the afterlife. Living as an embodied being is our one chance to take charge of our spiritual destiny and to determine our eternal future. Human beings have the capability of freedom and can choose the way they want to live, but whatever they choose consequences follow. As such, the collective mind of humanity, as a body of souls in the making, exists in a constant state of flux and change.

Many people are ignorant and uninformed regarding the spiritual reality and origin of both them and the material universe in which they live as embodied beings. So too are they unfamiliar with the esoteric knowledge that can provide them with information and instruction concerning spiritual growth, development and change of being. Bearing this in mind however, many sincere and well-meaning people do seek a spiritual, religious or psychospiritual pathway in life. They endeavour to live socially by observing ethical and humanistic principles, and they conduct themselves and their interactions with others in an upright and moral manner. Perhaps they follow the guidelines for religious behaviour and practice as set down in their relevant spiritual belief-systems and worldviews, and while these are noble and worthy aims they could be lacking because the inner or esoteric meaning of associated doctrine may not be known.

Dying to what we are to grow into what we want to be involves suffering. Learning the truth about ourselves can be difficult and painful but this is part of the process of self-cleansing. We must always remember that salvation through faith alone is a fallacy. To verbally confess our sins and wrongdoings before the Lord or one's God or deity does not automatically erase those things from our being. The Work is not an abstract doctrine of obscure ideas purporting to overnight enlightenment and religious freedom; when applied to our life it results in spiritual growth and change of being. Unless a person works on themselves with the specific intention of self-cleansing, awakening from psychological sleep, and growing and evolving a conscious soul, then nothing will change internally in their being. A belief in God is only part of the process and a 'head start', however far too many people think that is all that's needed. This situation produces individuals who are lopsided; they have heavenly aspirations yet do nothing about the chaotic state of their inner being.

Many people undergo valuable psychological and emotional changes which take place in the personality, but the being is not affected. We need *change of being*, this is crucial. This is where the Work offers us practical guidance and assistance because it teaches us how to transcend and pass through the personality to reach the essence; and faith, as *a process of becoming*, is a vital component of this teaching:

> The peculiar quality of faith lies in this idea, that life can only be understood and solved by the sense of something higher than Man as he is, and that Man has this possibility of becoming transformed and passing into entirely new meanings in regard to his life on earth. It is this peculiar quality that is the essence of faith and renders it utterly different from what we usually call belief. Faith, in fact, undermines all our ordinary and natural beliefs because it leads away from worldly belief and in a direction that can no longer be confirmed by natural belief and the evidence of sense. And for this reason it is defined as a seed in a man's mind – that is, it is the potentiality of a growth in a man's mind which cannot exist in him as long as he believes that life, as it is, is the end of Man, and not a means to something else. For if we imagine life is an end in itself and the only end, we cannot possess faith and do not wish it. But if the thought enters our minds that life cannot be an end in itself, but must be a means to another end – and often such thoughts occur to everyone – then, precisely in this

> moment of new thinking, there is the foreshadowing of faith. (Nicoll, 1984, p. 108).

Maurice Nicoll, who was a teacher of the work, personifies faith as having its beginnings in the mind as a seed, however while faith begins as a seed in the mind it needs love to grow. This is not *philadelphia* or brotherly love, this is love for God, *agape*, which is a different quality of love all together. The seed of faith has to be fed with love and affection for God and when this occurs the process of faith becomes a stepping-stone into new levels of spiritual growth. This occurs because we trust in the reality of that which, although unseen, is known to be indisputably present. What use is there in faith, if there is not love and affection for God as well?

How do we love God? We just do, but if we don't know how we can ask the Divine to help us. When I was a child, I felt that the whole world was saturated with something deeply profound, and so encompassing was this feeling that I not only felt that I was loved by this something which came from beyond the material world, but that I was a part of that 'something', that I was in it and it was in me. As young as I was, somehow, I understood that the universe didn't just love me; it loved everything that existed within it.

That love is still present, still as palpable now as it was then. But what I understand now that I didn't then, is that the material universe is not only an expression of love, it is the recipient of love. Gandhi defines it as "an indefinable mysterious Power that pervades everything", and he's right. It is indefinable and it does pervade everything, and it does transcend the senses. If we want to feel this, all we need to do is ask and Higher Life will respond. The golden rule is this; we need to *feel* God, not *think* God.

> If in the languages of men I speak, even of angels, and I have not love, I have become brass sounding

> or a cymbal tinkling. And if I have prophecies, and know all mysteries, and all knowledge, and if I have all faith, so as to move mountains, but have not love, I am nothing. But now remains these three things, faith, hope and love and the greater of these is love. (Paul the Apostle, 1 Corinthians 13:1-3, 13).

Work-on-oneself requires that we turn away from ourselves and turn toward the Divine as the source of all life, all being and all unity; that we reorientate ourselves. We are all a cacophony of 'I' states, a multiplicity, and though we regard ourselves as complete there is no unity within us, there is no Real I. Many people do sense within themselves that there is a profound presence that runs through the universe, and all of us in quiet moments at one time or another will feel or be aware of this. Our task is to bring this inner feeling into the open by drawing it closer to ourselves, so we can be more aware of it. This inner sense of a great presence, combined with the observation of the relationship and order of things, bring two poles of our being together. The deeper part of us feels presences, the upper part of ourselves sees relationships; we must try to get these two things operating simultaneously.

Many people demand proof of the existence of God and Higher Life, saying that then they will believe. Many people have expressed the sentiment to me, "How do you expect me to believe in something that I can't see? When I see it, then I will believe." These people, like the spirits who conversed with Swedenborg about miracles and faith, believe that only a miracle will confirm the reality of the Divine. Were such requests granted nothing would change; nobody needs proof of what they know. In John 20:29 Christ states so simply to Thomas, "Because you have seen Me, Thomas, you have believed. Blessed are the ones not seeing and believing."

The challenges of working on ourselves, although formidable, are not insurmountable. We need to know why we exist and how to make the most of our embodied or planetary life, and we need to understand the imperative of this. Our birth is an intentional arising, a promised coming forth of an embodied being into a world and existence the spiritual reality of which is obscured by our ignorance and our physical senses. Yet with us travels the hope and wish that we will see, that we will come to know the profundity and immensity into which we were born.

In concluding this chapter, I would like to quote American Christian pastor and author, Aiden Wilson Tozer (1897-1963) who wrote, "We need never shout across the spaces to an absent God. He is nearer than our own soul, closer than our most secret thoughts" (2015, p. 51). In the process of working on ourselves there will be many times when we may feel alone and unsupported; we are not. Tozer experienced what we all can. The Infinite transcends material existence as it transcends those who have no faith and no belief. It is there regardless, as it always has been and as it always will be and all it takes to experience it, is faith.

7

WHAT AM I?

A systems view of reality sees the world in terms of relationships and patterns of organization, and interactions, rather than as a composite of individual entities or structures existing in isolation, independently of one another. A systems view of the self therefore challenges the assumption that we exist only as alienated, isolated individuals in a hostile, or at best, indifferent, environment ... The self as an open living system is continually in flux ... In acknowledging the continual flux and exchange of energy between self and other, whether biological or psychological other, we grow to accept and develop more of our potential being ... From this perspective individuals do not exist as independent, isolated entities. In fact, they are continually changing, interacting, and joining with others in subjective exchange, in shared ideas and shared purpose.

Francis E. Vaughan, *Discovering Transpersonal Identity*, pp. 20-21, 1985.

One of the basic and obligatory requirements of the Work is the practise of self-observation. Because of carrying out this activity I learnt that I was a multiplicity. I came to understand that

I was many 'I's' as opposed to just one 'I', which is where my sense of self was usually centred. While I have always been able to see my behaviour and thoughts, I never knew that the act of seeing myself in such a way was self-observation in action; that understanding only came when I encountered the Work.

With this new understanding, it occurred to me that all of these 'I's' or 'me's' weren't me, they were in fact separate and individual little 'me's' or 'I's' that flowed through my mind seeking expression, and which *en masse* formed my personality. I understood that by identifying with them I not only became them, but that the picture I presented to myself and the world of who I believed I was, was inaccurate.

I saw that these 'I's' weren't real, that they were either an inflowing influence from an external source or were well-established 'I' responses that I was completely identified with but hadn't realised. With a shock I saw that I wasn't me as I understood myself to be, I was them, I was all of those 'I's'. I was fragmented, horribly so. I was not one, I was not whole, I was not undivided. I was *divided*, and I was *many*. Who I thought I was I wasn't at all, and with great dismay I realised that I was a stranger to myself.

It was at this time that I also understood how the personality formed, which appeared to exist as an external or 'outer' layer resting on top of something existing within; an interior spiritual pattern which acted as a metaphysical supportive framework. I remember thinking that the principle was like that personified by the Russian matryoshka or babushka nested dolls, where several dolls of decreasing size are contained within a single larger doll. I had had an experience of this 'layering' prior to entering the Work, however, and as for so many of these experiences, the context for understanding that was only provided after I had entered the School.

When I was fifteen years of age my mother took me to see a general practitioner. I had been experiencing peculiar 'turns' or 'episodes' and after confiding in her, and she being suitably alarmed on account, thought it in my best interest to seek medical advice. Some days later, I found myself sitting in the doctor's consulting room, watching him from across his desk. As we looked at one another I felt he wouldn't know what my problem was because it wasn't biomedical. My mother clearly didn't know, but at least she was trying to help. I told him what had been happening and afterward, as I knew he would, he looked at me with a puzzled expression on his face.

I suppose my account of suddenly feeling myself separate or come away from my physical body, such that I discernibly felt I was a body *in* a body and that there was now a discernible space between those two bodies which I constantly felt, didn't make much sense to him. Think of the old standard diving suits, the ones which consisted of the round goldfish bowel 'helmet', the heavy canvas suit and air hose which was connected to an air compressor on the surface – this is the perfect analogy. That's exactly what it was like. I felt like I was a body encased in another body which was enabling me to exist in a completely foreign environment, (and all these years later, it has never changed). I think it was my next statement that totally mystified him though. "I hear a strange sound, but it's not a sound, it's more a high-pitched frequency, and I feel like I'm being fine-tuned."

A minute or two passed, and then he leaned forward and said, "A lot of young girls, especially around your age experience this." Oh, perhaps I was wrong, perhaps he did know. "We call this, hysteria. It's very common and nothing to worry about, and it will pass." No, I wasn't wrong, he didn't know, but at least like my mother he had tried. It was Dr Groves who some years later provided the explanation. "Oh yes, this is the separation of the spiritual body from the physical body. This is something we all have to experience and it's a necessary part of your spiritual growth. The high-pitched frequency is the fine-tuning of your spiritual body, which will continue throughout your life" (*Dr Philip W. Groves, personal communication, 14 December 1985*). He was right. To this day I regularly experience 'fine-tuning' as I do the sensation of being a body inside a body and being aware of the space that exists in between the two.

After I entered the Work, I also came to understand the purpose of near-death and out-of-body experiences as well, which demonstrate the reality of ongoing existence beyond physical death. When the body is put off (termed 'death'), the spiritual body, which is the fundamental pattern of the individual, simply moves into the spiritual universe. Our individuality resides in our mind, which is our spirit. It's not lost at death, nor does it disintegrate, nor are any of its functions lost.

After death, and when we enter the spiritual universe, we live there with the same affections and loves that we lived with in the material universe; we live there (in this new realm) as we did here. The significant difference, however, is that we are now in the form of a non-material spiritual body rather than that of a material or corporeal physical body; we are disembodied. In 2001 I had an out-of-body experience which demonstrated this. I say out-of-body because although I was jettisoned into the spiritual universe, there was no one present who could confirm that I had physically died as the following account from my journal recounts:

> I had been bushwalking with my then husband in one of Sydney's national parks and for some obscure reason had decided to climb a tree. I've climbed trees since childhood, and scaled our household roof as well, with nothing untoward ever occurring. On this occasion I had wandered off by myself and it was a

huge smooth-barked eucalyptus tree which caught my eye. The tree having fallen over lay enticingly on its side jutting out over a rock ledge; perfect! Poised between the solidity and safety of the rock on one side and a sheer drop down several metres to trees below on the other, I was oblivious to the potential danger and immediately began clambering along the trunk which rose on a slight but steady incline.

It had been raining and I noticed that the trunk had an oily feel to it. Looking down toward it, my hair, which for once was not tied back, swept across my face obscuring my vision. Suddenly I lost my balance, the first time ever, and slipping around the trunk with my hands and holding on for dear life, I found myself dangling high above the ground.

The last thing I remember is feeling my hands slipping and thinking I was going to fall. I wasn't afraid; there wasn't time for fear because everything happened so fast. I felt my body drop, heard a loud 'thwack' sound (which I realised later was the sound of the impact of my head hitting the rock surface) a shift in awareness, and then to my great surprise I found myself in what appeared to be a great cavernous dark tunnel. I was in a different place, a different world, and I knew that I had left the world I usually lived in; I felt like I was in a different universe.

Time seemed distorted and I had a vague recollection that there had been an impact of some sort. I wondered to myself if I was dead. Was I? I didn't know. But I couldn't be because I was wondering if

I was, and if I was dead I wouldn't have the capacity to wonder about anything, would I? Wouldn't I be as nothing? But in the state I was in the realisation came at lightning speed; I was something, I existed. But I was different to what I was before, I was more real here. I was more 'me' here.

I didn't have a body like I did before because I was pure thought, but it was me, I was me. This me lived inside my body from where I came, but here it lived without that body.

I realised this because I looked down at myself 'seeing' nothing at all, and then I realised I was not in my physical body, I was outside it. I also understood that I was my mind, which is not physical; it is spiritual. Although I was in complete darkness and could not see, I felt that I was being held aloft at a great height, and though it seemed that I was floating, I was aware that a force or an 'intent' unknown to me was responsible for maintaining my elevation and my trajectory through this space. It was silent, serene, gentle, there was no noise of any kind, and I was completely alone.

In observing my thoughts, I saw that I had a curious non-attachment to what I had left behind; my husband, my son, my family, friends and possessions, in fact everything that had belonged to my life when I lived in my body. It was as though all those things simply didn't matter anymore, that they didn't count for anything because they weren't relevant in this place. Instead, other things belonged to me in this place, other things were more relevant, non-material things.

I understood intuitively that this was because a different rule applied here; it was a completely different existence and the things that had been of significance in my previous life were not of any use here. There was no arguing the matter; it's just the way things were, the way things are. It was like an unwritten code that for some peculiar reason I just knew. I could feel myself watching myself. How odd it was, to see my thoughts, to see myself seeing myself.

I thought of the people I knew, almost absentmindedly it seemed, and then I thought no more of them. Instead, my attention turned toward what it was that was keeping me elevated. I could feel that I was suspended at a great height; what was it that was keeping me aloft? Why didn't I fall? Up ahead I saw a pinpoint of light and knew that I was being moved toward it. I felt, not peaceful, but unconcerned about anything and it seemed that my mind was functioning differently. I was aware that the things that occupied it when in the other world, the world that I had come from, were of no consequence here. There was no worry, no preoccupation with self, no fear for tomorrow. People say, 'be in the present moment'; I was.

The light was becoming bigger as I got closer to it, and I could see that it was like a great circle through which I knew I had to pass. I was unconcerned about anything, and simply let whatever it was that was moving me, move me. There was no need to resist, I had no control over anything anyway, and then I passed through the light, gently and without incident, to find myself not moving any more but rather lying on my back looking up at several people standing around staring at me. I didn't know any of them, and none of them knew me, we were all strangers to one another.

One person, a young woman, wore denim jeans, a t-shirt and sneakers. Another, a man, was wearing a suit and hat circa 1940 while others wore what I would term regular clothes similar to those I would wear. Not a word was spoken by any of them, yet I knew their thoughts; all of them were thinking, "Who is this?" I had quite literally burst into their world, and they were all curious as to who I was. I shouldn't be here, it wasn't my time to be here, I knew it and they did too, and they were wondering "If it's not her time how is it that she's here? She hasn't died. She's still alive there."

I knew that I had entered the spiritual universe and had arrived in the world of spirits. I knew that I was looking at spiritual beings because I could feel gradations of qualities in each person. I felt a multitude of impressions bombard me and I understood them, could read them, and instantly knew the quality of each one of the people who had gathered around.

And then I saw something that startled me. At my feet stood two men garbed in Greco-Roman style clothing, brilliantly coloured; one of them captured my attention. The garments were layered, consisting of an under-tunic overlaid by a cloak. The cloak was orange while the tunic was red and belted at the waist with a plaited blue cord. But the colours weren't such as we know them. They shimmered and shone and glowed with an inner luminescence, and it seemed that I was looking at the very atoms of colour. The two men emanated an atmosphere that I could see. Like a great golden halo, it spread forth creating a sphere of light that sparkled around them, reminding me of the play of light reflecting across water droplets when a sun-shower occurs.

Behind them I saw several dark, grotesque, and distorted shapes. I understood instantly that these were infernal beings and became alarmed. I could feel their evil intent and knew that they wished to harm me, because that was their affection. I saw that the sphere created by the emanation of the two men wearing the luminescent robes was preventing these infernal beings from being able to come close to me and I understood that by their presence they were keeping me from harm. The dark beings could not tolerate the purity of the light which was cast forth, and so I was spared their intent.

In the next breath so to speak, I found myself being propelled over a landscape which shifted and changed beneath me. Brilliant colours flashed past, and voices, conversations, events occurring in people's lives, all of this I was privy to. It was as though I was being shown a collage of the lives of a multitude of other people, but I don't know why. Suddenly I was aware of a pulling sensation, a 'pulling away from', and a male voice, the voice of my husband. With a great whoosh coupled with the sensation of being drawn back into myself I opened my eyes and found him leaning over me; his voice had brought me back from where I had been.

And then confusion, disorientation, a time differentiation between there and here. Here, according to my husband I had been 'out' for only a few minutes yet there it seemed that eons had passed. I had travelled to another place and journeyed in a realm where time was not measured or quantified in the way that it is here. I had experienced two time-realms which coexisted simultaneously, as I had two worlds which coexisted simultaneously. The void I had traversed had been the thoroughfare between the two and had acted as a portal, of that I was certain.

I felt as though I had been on a journey to another world which in some ways was not so different to this one. There is good and evil there too, as there is here. There is an assortment of people there as there is here. What is different however, and fundamentally so, is how 'time' is experienced.

In our embodied existence time is measured and quantified, and it is well known that we live in a time-based universe. But in that other place, our perception and experience of time is completely different. There, time as we know it has no beginning and no end because it is not in operation, and the passing of our lives and existence isn't measurable because it is our state-of-being which governs our existence; we don't age because we are immaterial, we *are*. Corporeal existence is governed by decay, degeneration, and

> atrophy, but different laws appear to govern existence there and because of that I was aware of being somewhere which stands outside of time. In that place, I knew that what I had experienced was eternity.
>
> Somehow we managed to hike out of the national park, which took several hours. I collapsed several times almost blacking out. My head was pounding, and my body ached. We made it to a hospital where I was admitted into emergency. The right side of my body, especially my face, had borne the brunt of the impact and I had extensive bruising, lacerations, and a concussion. However, to the amazement of the attending physicians, not one bone was broken or fractured, and I had suffered no internal damage. I had fallen from the tree onto a rock platform some three metres below, after which I had somehow rolled down onto another platform some two metres below that. Somehow, I had narrowly escaped falling directly onto the trees below which could have been potentially fatal. (*Authors journal notes, September 2001*).

Several unusual events followed this, both during the hours I spent in the hospital and over the next couple of weeks, and it took some time for me to fully come to terms with what had happened. This was a deeply profound event with cosmological implications, and it was important to understand the many layers of information embedded in the experience. There is a persistent anomaly associated with what happened however, which is that I don't know how I didn't fall onto the trees which were situated to my right as I was hanging from the tree trunk; I remember that very clearly. The right side of my body and face had borne the impact of the fall, yet the rock ledge that I landed on was to my left; to this day I still don't know how I survived the fall.

Prior to this experience I had had no real interest in out-of-body or near-death accounts. I had had unusual dream experiences as a child, but what happened during the bushwalk occurred during the day when I was awake and fully aware of myself and my surroundings. In reflecting on the event, I could see several correlations of the experience with accounts compiled by Emanuel Swedenborg in his five volume *The Spiritual Diary* (SWEDENBORG SOCIETY, 1883-1902/1977). The volumes describe the spiritual universe and its inhabitants as being not dissimilar in appearance and quality to those of the material universe, and in his work *Heaven and its Wonders and Hell* (SWEDENBORG SOCIETY, (Inc.), 1758/1937) in which amongst other things, he discusses the intermediary World of Spirits.

There was something else as well. In *Heaven and Hell*, Swedenborg wrote that when a man dies, he only passes from one world to another, and that on this account, "Death in the Word in its internal sense, signifies resurrection, and the continuation of life" (1758/1937, p. 332). The correlation between what occurred to me with what Swedenborg had written was unmistakeable, hence the educational aspect of these types of events which countless people have experienced. 'Death' as an event corresponds to the action of the separation of the spirit from the body and 'resurrection' is entry of the spirit into the spiritual universe and continuance of life.

Returning to the revelation of my multiple 'I's', and the theme of this chapter, the realisation that I was a stranger to myself was quite startling and one I hadn't previously encountered. Who was I if I wasn't me? Who was I if I wasn't one but many? The Work had taught me that I was a conglomerate of 'I's' and a kind of unthinking mechanical machine that habitually and consistently reacted in a particular way to a given stimulus. I realised that I existed but that

I didn't really live, and that I made my way through life psychospiritually asleep and ignorant of the cosmological nature, cause, and source of my existence. While I had learnt in my childhood that life was multidimensional, it wasn't until this moment in time that I began to understand both the significance and implications of that.

My journals recorded my life experiences and in rereading their content I saw that I had compiled notes of how I had reacted to whatever it was that was taking place in my life at the time. On one page I was troubled, on another musing on the meaning and significance of a dream, and on another happy. It was a subjective record of who I believed was me, of who I understood 'me' to be. But this person wasn't me. The person I thought I was, was nothing more than a machine habitually reacting to whatever came its way. And knowing this was extraordinarily difficult to come to terms with.

To understand, I read and re-read my Work notes and material and pondered on all that I knew. While my spiritual experiences had provided an other-worldly or non-material frame of reference for my being in the world and for understanding that life and human beings were multidimensional, it was the Work that had taught me that I wasn't one 'I' but many and that though a mechanical machine, I could become conscious. But despite this I felt that the Work hadn't taught me *who* I was.

Who am I? became a burning question. It incessantly occupied my attention for weeks. I thought about it at work, while I was on the bus, before I went to sleep at night and when I woke in the morning. I thought about it when I was socialising with friends as I did when I was enjoying my own company. I thought about it while I was watching television as I did when shopping and even when reading it intruded into my thoughts. And then one day the answer came.

It was a Saturday afternoon, and I was at my parent's home sitting in a chair on the backyard verandah transcribing my shorthand lecture notes and typing them up, as I frequently used to do. The Summer sun beat down on my back and as I had been typing for some time my mind had become drowsy and relatively still. And then suddenly the answer to my burning question came to me like a bolt of lightning from out of the blue. Completely unexpectedly, it burst into my mind filling it with understanding which was so encapsulating that I was almost catapulted out of my chair. It wasn't *who* I was, it was *what* I was. Eureka!

In that instant I suddenly saw dimensions of my being that I previously hadn't been aware of, one of which was the possibility of becoming something other than *what I was at that time.* This was a fundamental psychospiritual shift in how I perceived myself and an important stepping-stone in my spiritual growth.

> *Man has no permanent and unchangeable I.* Every thought, every mood, every desire, every sensation, says 'I'. And in each case it seems to be taken for granted that this I belongs to the *Whole*, to the whole man, and that a thought, a desire, or an aversion is expressed by this Whole … Man has no individual I. But there are, instead, hundreds and thousands of separate small I's, very often entirely unknown to one another, never coming into contact, or, on the contrary, hostile to each other, mutually exclusive and incompatible … Each minute, each moment, man is saying or thinking 'I'. And each time his I is different. Just now it was a thought, now it is a desire, now a sensation, now another thought, and so on, endlessly … The alternation of I's, their continual obvious struggle for supremacy, is controlled by accidental external influences. Warmth, sunshine, fine weather, immediately call up a whole group of 'I's. Cold, fog, rain, call up another group

> of I's, other associations, other feelings, other actions. (Gurdjieff in Ouspensky, 1949, p. 59).

While the truth of Gurdjieff's words played out in my own life, and be that as it may, our situation is not hopeless. Yes, we are a rabble-like multitude however the gathering together of several better quality 'I's' of like-mind establishes within a person what is termed an 'oikonomos'. From the Greek οἰκονόμος, ου, ὁ (*oiko,* 'house' and *nomos,* 'rule, law'), the term translates to 'manager' or 'housekeeper' or someone who has the care of the household in the stead of its master. The allegory of the Deputy Steward (which is found in certain esoteric teachings and parables) is such that an individual's internal states are represented as a household in a state of complete disarray. The cause of the disarray is due to the household's many servants, who being left to their own devices and with no overseer to maintain order and harmony, have created internal chaos.

Sometimes certain 'I's' with perhaps a little more wisdom or who possibly because of certain life experiences know or are aware of this internal disorder, will band together in an effort to take charge. These more intelligent 'I's', known as the Deputy Steward, work in harmony with one another amidst the chaos that constitutes the internal and external life of the person. "Deputy Steward is exactly like the formation of a group in the Work, only it is internal, invisible, *in you*, and not outside you as is a visible group of people" (Nicoll, 1996, Vol. 5, p. 1640). If this cluster of 'I's' develop an affection for the Work, then self-evolution and a new way of being in the world is possible. Observing 'I' plays a critical role in this scenario because it is around this that these more intelligent 'I's' cluster. If observing 'I' is not present, then the Deputy Steward will not form, and self-evolution will not commence.

> Observing 'I' collects round it 'I's that can work and understand the work. They form a small group of 'I's

called Deputy-Steward, which have to struggle and fight, not only with False Personality but with undeveloped essence. If Deputy-Steward, in spite of endless failures, becomes strong enough, "Steward" draws near. "Steward" belongs to something above man. It comes at first in flashes and often when it draws near, people have great difficulties either externally or in struggles with negative states in the form of illness, etc. "Steward" comes from a different level. To receive, as it were, "Steward", a man must undergo a new setting of himself, a new ordering of his mind, or even brain cells. But this always takes place in the way best for the individual and can be endured. The work is to get in touch with higher centres. But they work in their own way so changes have to take place in a man. A man cannot produce these changes himself for he knows nothing of the new connections necessary. It is through his personal work and the struggle of Deputy-Steward in him that they are brought about—that is, what is trying to enter from above a man brings this about when the conditions are right. Once brought about, the man is a different man. His feeling of 'I' is different. His ideas and thoughts, his reasoning and his actions are different. He has undergone the self-evolution latent in him. He is "born again" as the phrase in the Gospels puts it. But all this is impossible unless a man establishes Observing 'I' to begin with and has the help of the work, via the understanding of it for himself, which means the clustering of other 'I's in him round Observing 'I', so that a small band of 'I's is formed in the chaos of his inner life called "Deputy-Steward." (Nicoll, 1985, pp. 37-38).

One of the requirements of the Work is that we 'die before we die', meaning that we die to erroneous beliefs and ridiculous assumptions and notions about Higher Life, ourselves, and the cosmological nature of reality. With the death of these attitudes a new understanding is provided by Higher Life which instructs the mind and spiritualises it. As Nicoll observes, we are born again.

The spiritual universe is a multilayered multidimensional structure. The study of the universal language of correspondences helps us to understand the mirrored relationship between the macrocosm and the microcosm in that all things in the natural or physical universe, including the human body, are a hieroglyph representing the spiritual processes and creative activities that brought them into being. The microcosmic reality of man as a reflection of the macrocosm is a fundamental aspect of Work teachings, and a principle which has brought forth the dictum 'As above, so below' and, 'As within, so without'.

Everything around us has a spiritual pattern; everything is a reflection or correspondence of something else which has come from a higher level of reality. Many people believe it important to know who they are and seek, as I did, the answer to that question. By growing into an understanding of *what* we are, we begin to see the illusory nature of the self and how it unknowingly distorts reality. The material world, the universe, and all visible creation, including mankind, is the manifestation and end-effect of a huge conatus, a consciring if you like of cosmic energies and patterns from the Divine mind. Each human being is the end-effect of the totality of this conatus, each human being is a miniature universe of equal and dazzling complexity.

When I understood the difference between *who* I was and *what* I was, I realised that I had crossed a gap in the octave of the scale of inner development. I experienced a quantum leap in my being because in the process of understanding that difference I had altered the internal structure of my being; my being had grown, was changed, was different. That moment was pivotal in my understanding of the nature and importance of the Work and of what its cosmic purpose is.

Understanding is a process which entails much more than simply remembering something about something else; it is the transformation of information into knowledge. Knowledge resides in our being in a space where the living actuality of what is being understood is felt. Our feelings bring our thoughts about a thing alive. They confirm and clarify what we *think* is right or wrong to what *is* right or wrong. The moment of understanding what we are is a life-changing one for the individual because it introduces into the mind the possibility of the process of spiritual evolution.

When I crossed that gap in the scale of octaves, I grasped the complexity of my being from the perspective of *what* it was. I saw that I *can* become something, not in a worldly sense, and that to do so I had to die to the understanding of myself as I did of 'reality' so the seeds of possibility and potential could grow into all that I could be. The Work is not given lightly, it is entrusted, and it must be held safely and carefully by the individual. It affects the spiritual regeneration and reformation of a human being, and it demands from us the birth and nurturing of a fully conscious soul. It is offered to all those who truly seek and yearn for a more meaningful way in which to live and which. All this I understood in that one instant, in that one quantum leap which carried me to a higher level of being as I sat in the sun all those years ago.

8
INTENTIONAL EFFORT

What the human can become is a wonderful question. It is related to our spiritual potential but also to our will to make something of ourselves in our evolutionary journey. All great religions and enduring forms of spirituality are eloquent manifestations of man's will to make something more of himself than an ordinary bread-eater.

Henryk Skolimowski, *A Sacred Place to Dwell: Living with reverence upon the earth*, p. 13, 2002.

In our daily lives we're required to perform tasks and duties which require certain efforts. These efforts are necessary however we become so accustomed to making these efforts that we forget about the intentional effort required for our Second Education, which is quite different. Undertaking work-on-oneself is difficult and one of the reasons so few people choose to commence the Work is because they find it too difficult to embrace. However, though the Work is difficult it is not impossible. The aim of the Work is soul growth and evolution, and if this were not possible it would never have been given to mankind.

Work-on-oneself requires intentional effort and in the struggle against oneself and toward the Divine and Higher Life we must be stubbornly persistent. We must struggle against our negative states with determination. We must struggle to awaken from the sleep-state, and we must struggle against the harmful influences that at times threaten to overwhelm us. There are many types of intentional effort that we utilise when working on ourselves, some of which include the following:

1. The struggle against faults, necessitated after a thorough exploration and self-examination of our being.
2. The organisation of efforts to change ourselves, accompanied by a thorough inner-searching.
3. The assimilation and digestion of truths into the understanding.
4. The entry of truths into the will.

Time and again we are required to carry out the activities of self-observation, non-identification and self-remembering and this to the extent that they become a way of life, like breathing. Time and again we are required to digest the impressions that enter us. Time and again we are required to ponder, think deeply, and contemplate the timeless truths communicated to us by the Work. Time and again we are required to reflect on what our purpose and aim are. And above all, we must remember that without effort nothing is achieved. This notion of intentional effort is captured in the following quote by Indian writer and author Yashpal Jain (1912-2000):

> The greatest handicap to the search within is that we do not know the goal of life. If we start on a journey without a clear idea of our destination, we wander aimlessly. The first essential is to have a clear perception of life and its aims. The right knowledge is followed by right faith or determination that we will not rest until the goal is reached. Our scriptures say, "Ultishat, jagrai, prapya baranni bodhat," or "Get up and stop not until you have reached your destination."... The

> determination to reach the goal has to be followed by firm action. (Groff, 1984, pp. 112-113).

Unbeknownst to us, our birth and subsequent life as a planetary dweller herald to the spiritual universe the possibility of the entrance to that realm, in time, of a conscious man or woman. For those of us who choose to work on ourselves we must do so with eternity in mind. What is our spiritual destination? What is, as Jain observes, the goal of life? What eternal future do we wish for ourselves? How do we get there? And once we've arrived, then what?

The journey and process of answering those questions reveals how Beyond the Fourth Way complements the Fourth Way, because once we've awoken from the sleep-state we can focus on building our relationship with the Divine and Higher Life, and it is at this junction that Beyond the Fourth Way offers practical guidance.

While I was editing this chapter, I was overcome on one occasion by a sense of despair regarding certain aspects of my life situation. I was driving to work, it was early morning, and completely identified with the state I felt tears well in my eyes. I had switched on the radio but even the silly jokes and stupid antics of the radio jocks couldn't bring a smile to my face. As I began to sink into the state (because I was identifying with it) a sense of great calm unexpectedly come over me and my chest felt a little peculiar.

The state of despair vanished instantly, and in that moment, I *knew* that God wasn't just with me, He was *in* me. Out loud I spoke the words, "Father, thou art with me" and as I did each word reverberated within my being and a profound sense of awe and love for God rose from within and then I did cry, not tears of self-pity as before, but tears of love. With eyes inside my eyes, I saw my ribcage slowly spread apart to reveal a great yellow sun nestled within its cavity. I felt that I wasn't alone, that I had never been alone because God was constantly present *within* me as a living, loving Being.

I understood then that with the advent of my birth God had always been within me, anchored, and I understood too that the Divine wasn't just present in me, the Divine was present in every single human being; I felt it. All of us, whether we know it or not, whether we experience it or not, whether we believe it or not are anchored to and in God, and vice versa; we have to be, or else we couldn't exist. We are all of us the recipients of life, and we all come from the same spiritual source, are composed of the same spiritual substances, and are ultimately one race, the human race. When I had this experience, something psychologically shifted and I saw fundamental aspects of what was real and what was not as I did the universal truth of our spiritual origin and existence.

When I was a child, I had a powerful experience which demonstrated this sense of universal oneness and connection in an unexpected manner. An annual event which was the equivalent of the Easter Show in Sydney, was held each year in Boroko. This event was a favourite on everyone's social calendar and Indigenous people's from all over Papua New Guinea gathered together to participate in grand displays of traditional tribal dancing, drumming, singing and other events. Indigenous foods could be sampled and hand-crafted billums, carvings, hunting tools and other items were available for purchase. As I wandered through the attractions, I became separated from my parents in the crowd and rather than being fearful at the separation, found myself instead drawn to the sound of men's voices singing.

Making my way through the crowd of legs I pushed through until I managed to stand in front of everyone. Before me stood a line of men in full tribal array singing and stomping their feet. Their eyes looked strange, and I knew that though their bodies were there, they themselves were somewhere else. The line of twenty or so moved and shuffled forward in complete unison; it was a fantastic sight and utterly mesmerising and I stood still, unwilling to move from where I was standing even though I knew my parents would no doubt be worrying.

As I watched them, I suddenly felt the world fold in on itself. I could feel myself pulling away and retracting from the confines of my physical body, and then I found myself in another place; perhaps it was an altered state of consciousness, but in truth I don't know. In this place I felt understood, and knew instantly that all are one, that there is no division between people because all are composed of the same substance, the same type of matter. But it wasn't matter, it wasn't a solid substance because it was something else entirely that I didn't know the name of, but I felt it. I understood that there was no separation between the tribesmen and me because on the inside I was the same as them, that is, the spiritual pattern of our being was the same. I understood too, that because we were composed of the same spiritual substance and had the same point of origin, that in a sense I was them and they were me; we were together, yet separate, each other, yet one. I understood this because I *felt* it, but to add to this, I felt that we were one in the substance that is God.

I felt myself dissolve into this oneness until all that was left was 'not me', not the physical me that is. 'I' was something else entirely because I had dissolved into the origin of everything. A great blue-white light exploded in my mind and in front of my eyes I saw an endless sea of stars, and then suddenly it was over. I was back; I was 'me' again. I felt the hot sun on the top of my head and body, could taste the dirt in my mouth which had been kicked up by moving feet, and then rather unceremoniously the hard grasp of my mother's hand on my small arm as she pulled me away from the tribesmen with an admonishment.

Thinking as a child, what I knew in that moment was that underneath our bodies, which were all different on the outside, we were all the same and though we might look different on the outside on the inside there was no difference at all. It was the inside that we really were, not the outside. There is a correlation here with what Sir James Jeans (1877-1946) English physicist, astronomer and mathematician has to say about this state of 'oneness'.

Positioning his sense of oneness within a physics context, Jeans discusses the particle-picture concept in which each particle and photon is a distinct entity existing independently. When progressing further toward reality however, then the wave-picture is encountered. Here, because their separate individualities merge the photons are no longer independent individuals because they are now members of a single organisation or whole, a beam of light. The same principle applies to electrons. In the wave-picture they too lose their individuality and merge to become an electrical current. On a spiritual level, I feel the same applies to us.

> When we view ourselves in space and time, our consciousnesses are obviously the separate individuals of a particle-picture, but when we pass beyond space and time, they may perhaps form ingredients of a single continuous stream of life. As it is with light and electricity, so it may be with life; the phenomena may be individuals carrying on separate existences in space and time, while in the deeper reality beyond space and time we may all be members of one body. (Jeans, 1981, p. 204).

This merging into a unified whole, where all become indivisible within what Jeans' describes as a single continuous stream of life is what the experience previously recounted about the tribesmen demonstrates. This is another higher essence experience, and its value is in its disclosure of the spiritual reality and origin of the material world from the perspective of the Divine. The material universe and everything in it is the end-effect of a "vast series of interconnected processes reaching from world to world, and finally coming into visible and tangible manifestation" (Groves, 2000, p. 9). As embodied beings we too are an end-effect of that self-same creative process, and like the material universe not only do we have for our origin Absolute Life but are a part of "a vaster consciousness that links him [man] with all beings in all spaces and at all times" (Groves, 1996c, p. 100).

As an adult I have reflected and pondered deeply on all my childhood experiences, only some of which are included in this book, and know with complete certainty and assuredness that spiritually we all come from God, and that we are all one in God. This is what we are waking up *for*, for God, and this is why we work on ourselves. Irrespective of our physical manifestation, our differing personalities, what culture or society we are born into or might adopt during our lifetime, or how we define our gender and sense of self, human beings all originate from the same spiritual source and are all composed of the same spiritual substance; we all have the same spiritual pattern. And whether we think it, know it, or believe it, we are all spiritual beings in a physical body. This universal principle applies to the physical manifestation of the material universe.

> Science provides us with an exciting and very detailed external view of the universe, and the spiritual procedures of Ancient Wisdom make possible an extensive internal view of the universe. In order to get a wholesome image of reality we require both views and must see how they dovetail together. (Groves, 2000, p. 9).

The interactive relationship between science and spirit, in the context to which Groves refers is an important one, and to understand this more fully it is necessary to come to terms with the fact that "physical things are derived from spiritual precursors and are the hieroglyphs or symbolic representations of spiritual realities" (Groves, 2000, p. 11). The spiritual experience of oneness in the 'origin of everything experience' which occurred to me as a child finds its' parallel with the scientific view of Jeans' wave and particle-picture, and graphically demonstrates the wholesome image of reality which Groves refers to.

When we become emotionally involved with the Work, when we *feel* rather than intellectualise about its value, then the emotional parts of our centres become activated, and we start to function from higher levels of our being. Experiencing higher essence states is an indication of this. This is where Beyond the Fourth Way complements the Fourth Way so well; the Fourth Way assists us in awakening from the sleep-state while Beyond the Fourth Way teaches us about the spiritual foundation of reality and where we are going in the universe. Once we are emotionally involved with the Work the will, which is the instrument of love, also becomes activated and we interact and engage with the Work in a fresh and new way.

The will is "everything in a person which loves, likes, has affection for, desires, wishes, wants, craves, lusts after, and needs. It is the internal driving power which makes man what he is, and the nature of a man's will determine his quality, as well as the quality of his thinking" (Groves, 1996c, p. 58). Groves further elucidates the importance of the will and the understanding in the Work in that "whereas the understanding relates to thought, knowledge and truth, it is the will which relates to perception, feeling, love and affection" (Groves, 1996c, p. 58).

We must be careful not to become a 'knowledge-machine' (Groves, 1998b, p. 19) where all the 'right' aspects of working on ourselves are remembered and all the 'right' activities are carried out in a systematic and repetitive manner at certain times of the day or month. And it is here that the will working together with the understanding can assist us; we understand the knowledge of the Work and because we love the truth of that knowledge "from Love arises the complete and perfect knowing of the inward nature of all things" (Groves, 1996c, p. 58). Above all, we must remember that we cannot remember ourselves unless we feel that there is something higher than us. Unless we feel this, "our self-remembering will always lead us into false personality" (Groves, 1997d, p. 14).

It is particularly difficult to explain the feelings that accompany spiritual experiences, such as those included in this book, and my thoughts are that it's virtually impossible to do so. The way I experience them is uniquely my own and when they take place, I know with an inner certainty that something of a profound nature is occurring. For example, one of the things I realised many years ago, was that even though it felt like I was getting myself out of my negative states I knew that the power or activating energy to actually separate myself from them didn't come from me. I had to have the will to want to step out of them, to want to change, that was certain, but the enabling of that, that came from Higher Life.

The following analogy might help; a light bulb (me) can only be powered into existence by electricity (Higher Life). In the same manner, Higher Life has always acted in response to my intent, and this is another reason why I have often felt that the spiritual experiences that have occurred in my life have been done *to* me, rather than *by* me. When change occurred, as it always has, I knew it was never me who eventuated it; it was always God who did the real Work. I am one small human being amongst billions of others, and though I work on myself with all my will, I am always aware of my nothingness and my inability 'to do'. In the midst of all that is dark and unworthy, and all that I have inherited phylogenetically from my ancestors (in that within my being reside their deeds' and wishings' and yearnings'), I know that only God has the power to raise me from myself, which was demonstrated by the experience which occurred when I was driving my car.

Esotericism gives us the following instructions: Sleeping man must awaken (Awakening); Awakened man must die (Death of old states); New man must be born (Birth of new states). "The creation of a New Man or Woman is the outcome of prolonged, conscious, intentional efforts in accordance with certain principles and teachings" (Groves, 1998c, p. 14). The process of awakening, the death of old states and the birth of new ones require effort, and the determination to reach any goal must be followed by committed action. This is our cosmic duty and one that affects all of us, and we need to know what to do and how to do it.

We live from our personality and are asleep to the nature of our inner spiritual reality and of Higher Life. When we work on ourselves, and as we slowly awaken from the sleep-state, we begin to understand that we need to 'die' to many of those elements that up until that moment in time have significantly contributed to shaping who and what we are; our negative states, our mechanical reactions and our ignorance. In dying to those aspects of ourselves we detach from them and in time are born anew in the sound light of esoteric truth and doctrine. This is a conscious person, a person who has become part of the dynamic activity of soul growth and self-evolution, a person who is fully conscious of their cosmic duty and responsibility about themselves, those around them, and Higher Life.

In making the sort of effort required in the Work, we need to compel ourselves. What exactly does this mean? The etymology of the word 'compel' provides practical insight. The word is derived from the Latin 'pulsate', 'to throb', which in turn is derived from the Latin 'Pellere', meaning 'to drive', which in turn is derived from the

Aryan root SPAR, struggle (Skeat, 1993, pp. 376, 597). The word 'effort' is derived from an old-French word 'force', 'strength', which in turn is derived from the Latin 'fortis', 'strong', in turn derived from the Aryan root DHAR, 'sustain', which itself is a composite of two other Aryan root words DAR, 'to do' and DAL, 'to see' (Skeat, 1993, pp. 158, 590-591).

These root meanings are diagnostic because they provide an insight into some of the elements involved when we compel ourselves to make an effort, which is a particular type of activity. Before any effort can be made however, we need to see what needs to be done, then with strength of purpose we begin. The outer self is lazy and usually thinks up an excuse for not doing this and tends to put off to tomorrow what it decides cannot be done today. And this is the very moment that we need to take charge of the situation and *compel* ourselves.

We all have the freedom to continue the effort or to give up, the choice is ours. Making the decision to continue involves a struggle because the inner self is trying to instruct and guide the unruly outer self, something the outer self strongly resents. When we make the decision to carry on with our effort, we initiate a kind of internal pulse or rhythm. The *Australian Concise Oxford Dictionary* (Oxford University Press, 1992, p. 1210) defines the meaning of throb, from which the word compel derives as 'persistent rhythm'. This is a nice summation of the process of compelling oneself to make efforts, because when an effort is made it is like an ongoing internal rhythm. The rhythm is the action of the effort combined with the persistence of our strength of purpose and commitment to ourselves.

For what purpose do we undertake these efforts? When we are conceived, we are endowed with an essence which is spiritual and eternal, and which has to grow, develop and be perfected. The essence must take into itself certain energies and experiences from the world that it is born into so that it can become a conscious soul. Being endowed with a physical body and various senses enable us to interact with the world around us. This results in countless impressions being obtained from the world which are not only never transformed, they form layers of undigested material.

These layers of material constitute the personality which is the outermost part of the psyche. Every day that we live we add more material, and as we go through life our personality becomes thicker and denser and so it becomes harder to reach our own real self. The personality contains undigested impressions, all sorts of personality reactions, fixed habits and a great deal of information, and the Work teaches that we must digest these impressions; we must go over these things and work with them. What happens in time is that our centre of gravity comes to rest in the personality, which is external, rather than in the essence which is internal.

As a result, we have quite literally turned away from the essence and what we need to do, is rotate within ourselves away from the personality and back toward the essence; we need to turn around inside ourselves. The goal is to reach the real part of ourselves which never dies. Once we achieve this 'turning around' personality, in time, will gradually be shed and we will find that how we live in the world and how we experience being in the world changes.

How do we reverse the polarity within ourselves? How do we wake the essence up so that *it* becomes active in our being rather than the personality, which should not only be passive, but which should serve the essence? Here again the Work provides practical guidance by way of an exercise. What we need to do is move psychologically in a new direction, and we do this by repeatedly visualising ourselves physically turning around inside ourselves so that we face from the previous outward direction to an inward direction toward the essence. We need to do this repeatedly, and to do it repeatedly requires intentional effort. This is what is known as a being-effort because the action of turning around inside ourselves literally helps to grow our being.

The transformation of the essence during planetary life into a fully formed conscious soul is our alchemical task in life. The soul has a destiny of its own, a destiny that is solely determined by the efforts made by us during our embodied life. Our first great effort in life is to become conscious, our second great effort is to grow and transform essence into our soul. Higher worlds are composed of conscious souls, and Higher Life consists of vast communities of souls, whom by their harmonious activity generate from themselves external communities of like-minded beings. Human beings place themselves through their efforts into one of these communities, be they heavenly or infernal, to share eternity with their spiritual kith and kin.

We must be careful when we enter the Work that the personality is not allowed to take its instruction and spiritual truths and deal with them in a superficial way; if this happens nothing will reach the essence (which is the nucleated soul). Spiritual truth *will* reach the essence. If we can learn to love spiritual truth it can make an inroad into the essence, which is hungry for this higher food. Food for the essence is of a special nature and is termed 'manna'. Revelation 2:17 provides the following insight:

> The [one] having an ear, hear what the Spirit says to the churches. To the [one] coming, I will give him to eat from the manna hidden. And I will give to him a stone white, and on the stone a name new being written, which no one knows except the [one] receiving [it].

The language of correspondences reveals the esoteric meaning of the parable as follows. 'The one coming' is the inner part of us and is the stage at which the essence has grown and learnt to rule the unruly personality. The essence must be strengthened by our intentional feeding of it with spiritual truth. Because of this intentional feeding, which is part of Work practice, the essence is strengthened and grows to the point where there occurs a life and death struggle between itself and the personality. At this point we must make extra efforts. Gradually, essence will reach the point of dominance and will become active; essence is the 'one coming'. The 'white stone' corresponds to celestial truth which enlightens the awakened essence. The stone contains a 'name new' (the name of a thing signifies the quality of that thing). A person who grows strong in essence receives a new quality in their essence.

The idea of effort is mentioned in many sacred writings and texts and its importance cannot be overlooked. For example, in the beautiful English translation of the *Song Celestial* (otherwise known as the Bhagavad-Gita) (Theosophical Publishing House, 1975) by English poet and journalist Sir Edwin Arnold (1832-1904), the following excerpt is taken from Krishna's response (in the Karma-Yog or the Book of Virtue in Works) to a question posed by a troubled Arjuna. The Song Celestial is a Sanskrit poem which comprises part of the Mahabharata, and its plot is in fact a dialogue which occurs between Krishna, the Supreme Deity disguised as a charioteer and Prince Arjuna, brother of King Yudhisthira. Both charioteer and prince are drawn up alongside one another waiting for battle to commence between the two opposing armies of the Kauravas and the Pandavas, and while waiting, engage in a meaningful dialogue about the importance of spiritual work.

Their discourse encapsulates and highlights the necessity and difference of inner spiritual work as opposed to external outer work, as Krishna states, "There is a task of holiness to do, unlike world-binding toil, which bindeth not the faithful soul; such earthly duty do free from desire, and thou shalt well perform thy heavenly purpose" (1975, p. 25). Yes, there is a task of holiness to do, and yes, it is unlike world-binding toil. The task of holiness is soul evolution and the toil required, which is very different to world-binding toil, is intentional effort.

It is important while making this intentional effort that we understand what we are working for, and this is what Beyond the Fourth Way teaches us. We are building for ourselves our spiritual future. What we make of ourselves while living as an embodied being in the material universe is what we take with us into the spiritual universe. Through our effort, and with vivification by God, we grow our soul (which is essence transformed). Death is the great leveller and what we make of ourselves is what we take with us into eternity. Death strips away titles, possessions, wealth, status and position in society and it is a useful exercise to occasionally review our life and ponder on what we think we have made of ourselves.

We get from the Work what we put into the Work. The responsibility for our spiritual development rests squarely on our shoulders and the intentional effort we make. The Work is difficult, yes, but it is also a journey of discovery, awe and wonder, of ineffable joy and at times sublime ecstasy. The material universe in which we live as embodied beings provides the perfect backdrop for our transformation from sleeping machines to active participants in the spiritual universe. All it takes on our part, is effort.

9

A WORD ABOUT DEATH

> And if the tree falls to the south or if to the north,
> where falls the tree there in the place shall it be.
>
> Ecclesiastes 11:3.

Death and the afterlife have always beckoned to me, not in a morbid self-obsessed way, but in a way that has always incited curiosity at its presence, and in an intelligent universe where it seems to me that nothing exists unless it has a purpose, I've often wondered what then is the purpose of death? If we are born to ultimately die, because we all do, why are we born at all? What is the sense in our being born?

As a child, I used to wonder why living things existed when all they seemed to do was eventually die or 'stop being'. Why did leaves wither on a tree? Why did the mangos that fell from the great branches of our backyard mango trees, which ensured they were always out of reach of little arms and hands, rot and eventually shrivel to a blackened pulp while they lay on the ground? Why did our family dog die? Why did the beetle I was keeping in my favourite beetle-keeping jar die?

As I observed the world around me with great intensity, I eventually concluded that what mattered was not *how* things died, what mattered was *that* they died. This bothered me then as it did for many years afterward, not because I was afraid to die but because it seemed to me that there was more purpose to our being in the world than simply being born, living, and then dying.

Death is generally regarded as the final act in the drama of our embodied lives and existence, however in a spiritual sense, death simply means resurrection into eternal life. Throughout this book constant references have been made to the fact that our spiritual destiny, our afterlife, is shaped and determined by the efforts we make while living our planetary or embodied life. Higher Life holds its breath, waiting, but that doesn't mean when we die we march on into a sublime celestial existence. It's up to us to create the spiritual future we want for ourselves here and now, and we do that by working on ourselves. Do we understand what we are? Do we understand why we are born? And most importantly, do we understand how precarious our cosmic situation really is? This principle was known in ancient times and can be summed up in the following parable:

> And if the tree falls to the south or if to the north, where falls the tree there in the place shall it be. (Ecclesiastes 11:3).

The language of correspondences assists us in understanding the parable's esoteric meaning in the following way. The term 'tree' corresponds to our being, specifically what we have made of our being, what we have spiritually become or not become. This is the totality of the lifetime of intentional effort we made when working on ourselves. The term 'falls' corresponds to physical death. The words 'there in the place shall it be' refer to the fact that what we have made of ourselves up until physical death remains so for eternity. The term 'there in the place' corresponds to our spiritual development and growth, which is representative of a particular level and spiritual society in the spiritual universe to which we will belong. 'Shall it be' is a warning that what we have or have not become, will remain as such.

The words from Ecclesiastes are a decisive and powerful communication from Higher Life which illustrate the importance of work-on-oneself from a cosmic perspective; we are what we make ourselves to be. We determine where and how we live in the afterlife; it really is that simple. There are similar exhortations in other sacred writings which are all as deeply meaningful and richly symbolic. For example, Kabir advises:

> O friend! Hope for Him whilst you live, know whilst you live, understand whilst you live: for in life deliverance abides. If your bonds be not broken whilst living, what hope of deliverance in death? It is but an empty dream, that the soul shall have union with Him because it has passed from the body. If He is found now, He is found then. If not, we do but go to dwell in death. If you have union now, you shall have it hereafter. (in Tagore, 1915, pp. 2-3).

In the First Khanda of the Eighth Prapathaka, 'Concerning the nature of the soul', in the Chandogya Upanishad we read:

> For, just as here on earth human beings follow along in subjection to command; of whatever object they are desirous, whether a realm or a part of a field, upon that they live dependent … Those who go hence without here having found the Soul (Atman) and those real desires (satya kama) – for them in all the worlds there is no freedom. But those who go hence having found here the Soul and those real desires – for them in all the worlds there is freedom. (Hume, 1931, p. 263).

The Gnostic texts also highlight the importance of working on ourselves now, as they do the sense of immediacy which accompany that task. In The Gospel of Thomas from the Nag Hammadi Library we read the following words purportedly spoken by Jesus to his disciples, "(59) Jesus said, take heed of the living one while you are alive, lest you die and seek to see him and be unable to do so" (Koester & Lambdin, 1978, p. 132). Again, the exhortation is clear. Know now, learn now, prepare now.

Planetary existence and our embodied lives provide us with the opportunity to work on ourselves. The intentional efforts made while in our embodied state are crucial to determining what sort of eternal spiritual future we create for ourselves. What we make of ourselves here remains so there, and once we arrive there, how we live is reflective of how we lived as planetary dwellers. The Work makes it very clear that a large part of our life should be devoted to active work-on-oneself in preparation for the afterlife. In fact, our early years should be spent in gathering esoteric knowledge which in our later years is digested and absorbed to bring about the required transformations necessary for the evolution of our soul.

In the examples provided from the sacred texts there is a repeating theme; do what must be done in this life to benefit oneself in the next life. In the Christian Bible there are several references to 'sowing' and 'reaping' (i.e., Job 4:8, Galatians 6:8, Luke 8:5). We need to sow the seeds of truth in our lives (which translates to 'understand and apply spiritual truth'), to reap later (which translates to the fruition of our efforts from working on ourselves). Our spiritual destiny is created while we live as embodied beings, and it is our actions and efforts that determine what that destiny will be.

The spiritual universe is a multilayered cosmological structure composed of innumerable societies of spiritual beings of different states and gradations of love and affection (see also the Koran where reference is similarly made to the "eight gardens of Paradise" and the "seven hells" which correspond to the same). Those of like-mind and like-love automatically gravitate toward one another to form vast communities. The nature and quality of the life we lived while embodied determines the type and quality of spiritual community we join when we enter the spiritual universe. This has profound implications that simply cannot be ignored. Human beings have a cosmic purpose that they need to wake up to as well as a spiritual destiny which is theirs by right of birth. But we all need to claim it.

The earth on which we live out our embodied existence is the realm of transformation and regeneration. Here we are provided with opportunities which enable us to take an active role in determining the nature of our eternal existence. Unfortunately, many individuals enter the spiritual universe in an untransformed and incomplete state which is the result of their lack of effort while living. It is the wish of the Divine that we all enjoy the bliss, harmony, and delight of the celestial regions of the spiritual universe, as it is that we live a life of use and eternal happiness. The crux of the matter is we have to wish for that too, and we have to make intentional efforts to actualise that wish, which we do by working on ourselves.

> How can we come to realise that we live in a participatory universe, that each of us, each minute part of us, is a node within a vast network of creative dynamics, unless we engage in practices that awaken our minds to the realities of such participation? (Spretnak, 1993, p. 22).

We do live in a participatory universe, and we are part of something profoundly vast and creatively intelligent. Existence in the better levels of the spiritual universe is not automatically granted. We have to work for it, and the harder we have worked on ourselves the greater the possibility of securing for ourselves an eternal afterlife that is loving, useful and cosmically fulfilling. If we haven't thought about eternity, we need to do so now. Physical death signals our entry into the timeless realm of deathless existence. In what condition do we enter? Do we enter with a being polluted with negative states and tendencies, an undeveloped essence and a soul which has not evolved? Or do we enter as a fully conscious Man or Woman, a harmonised being with an evolved soul and the wish to be of use to God and Higher Life? Remembering to occasionally ask ourselves the following question can help galvanise and refresh our efforts, "Am I in a fit state to enter the spiritual universe?"

Life has taught me that there is no death, not in the spiritual sense; there is only deathless existence as there are worlds within worlds. Life has also taught me, as it has others, that the event of death enables a human being to change their form. But there is more to it than that, and to state it so simply is to deny death's purpose simply as the doorway to our afterlife.

When we die, we enter the vastness of the spiritual universe, a world largely unexplored by most people. Sometimes our lives intersect with death, as in for example shared-death experiences or near-death experiences. These events teach us that we are more than our material selves and that our existence continues after death, albeit in different form. Such experiences can be deeply profound for the individual, creating psychological shifts and a reorganisation of their inner lives, their world views, and ultimately, how they live in the world or how they leave it.

Many people have said to me throughout my life that death is a mystery. Dying isn't, that's all too real, but the meaning of death and its purpose in our lives doesn't have to be a mystery and it won't be, I believe, if we can find our rhythm with it. How we do this is intensely personal, unique, and dependent on numerous factors all of which may interconnect with other facets of our lives, which themselves can be subject to influence and change at any time.

My journey toward finding my rhythm with death and the afterlife began in my childhood, a formative period during which one of the things I learnt was that I had the power to think for myself. I also learnt something else, which was that as young as I was, I knew that how I wanted to live in the world was the result of my own inner decisions; it was later in life that I learnt how those inner decisions were contrary to life's influences. And it's been that way ever since.

From the perspective of an afterlife and a useful universe, what is death's use and why would we contemplate it? It seems to me that when we contemplate death, by default we also contemplate life, and when we contemplate life, we eventually ask ourselves, "What am I living for?" Some people may never ask themselves that question, yet others who are discontent with life are inwardly driven because they *know* life can be more meaningful, because they *know* there must be something more to life than simply being born and then dying at some point. And so they search for a way, or a signpost which will point them in a direction which will lead to answers for them.

At the moment of our death we leave the material universe forever, as we do our earthly bodies. Do we enter the spiritual universe as incomplete beings with a poorly formed spiritual body, ignorant and ill-equipped to play a functional role in the maintenance of the spiritual universe? Or do we enter as a conscious being with a fully evolved soul and a well-developed spiritual body, and, as someone destined for joyful and active participation in celestial life? The answer to those questions, as is what determines the nature of our afterlife, is up to us.

10

A NEW WAY

> What secret do these men bear with them to the tomb? Why are they wondered at without being understood? Why are they acquainted with things of which others know nothing? Why do they conceal what all men burn to know?
>
> There is indeed a formidable secret … there is a science and a force … there is one sole, universal and imperishable doctrine, strong as the supreme reason, simple like all that is great, intelligible like all that is universally and absolutely true. This doctrine has been the parent of all others … the secret constitutes the science of "good and evil", and the secret of indefinite human progress is in that expression "the kingdom of heaven."
>
> Margaret Anderson, *The Fiery Fountains*, p. 242, 1969.

At one stage during the writing of this book I was momentarily overcome by doubt and a sense of inadequacy. And yet as I come to the end of this odyssey, I know that what has been written is only the briefest of introductions to the Work. There is so much esoteric teaching in the Work that a hundred books, even a thousand books couldn't come close to transmitting all that the Work is and besides, there are already excellent books in print which have been written by Fourth Way writers containing detailed explanations and accounts of many aspects of the Fourth Way.

This book was never intended to be a regurgitation of information about the Fourth Way already in print. My instruction was to put all the Work into a book, not the Fourth Way and not Beyond the Fourth Way, which is an important consideration. The Work, as it was taught to students in Dr Grove's School, encompasses the Fourth Way *and* Beyond the Fourth Way.

> For any person the primary thing is to acknowledge the Divine of the Lord, and to recognise that apart from Him there is no salvation. There are many mistaken beings who assume that Fourth-Way work by-passes the need for God, and that one can build a kesjan body and enter Higher Life without reference to religion. They are blind to the truth that genuine religion is the deepest and most necessary thing in human life, and that in its absence a person is a cosmic outsider. (Groves, 1998d, p. 7).

While the Fourth Way provides important instructions for how to awaken from the sleep-state it is Beyond the Fourth Way which teaches us about "the spiritual foundations of reality and the souls eternal prospects" (Groves, 1998f, p. 13), and while this book is firmly positioned within a context to 'know thyself', it is actually a synthesis of both lines of esoteric teaching and doctrine. The autobiographical content and reflective and narrative tone of this book demonstrates this synthesis, this meeting of the ways, which is woven into the title of this chapter. The Work, which encompasses them, comes from the Divine and is offered to all humanity. The challenge we face in the contemporary age, is to bring the two ways together so both are operating simultaneously.

The conversation during which the seed for this book was planted took place at what seems an eternity ago now, yet the moment is just as vivid and clear now as it was then. Since that time life has consisted of events, one after the other, many of which have involved prolonged periods of vastation, a painful process (for me) of inner purification and cleansing. The internal resilience and strength required to endure oft accompanying psychospiritual and physical pain was given by God when I asked for it, and in so doing I have been able to endure great suffering.

I know I have to work on myself, and I know I have to work hard, and perhaps, at least for me, what is significant is that this can be done in the midst of living an every-day life in the Western culture into which I was born. I almost feel embarrassed when I think back to the time when I thought that either travelling to Tibet or entering a religious Order of some sort was the only way to be 'spiritual'. We don't have to do that. Greek Orthodox theologian and author Panayiotis Nellas (1936-1986) agrees. "Moreover it becomes clear that the spiritual life is not an escape from the world but a transformation of it. It is not a change of place but a change of manner of existing and living" (1987, p. 148). In fact, the Work is perfectly suited to a busy external life because the circumstances and events which occur in our lives provide the very materials that we need to work on ourselves.

We all stand poised between two great realms of positive and negative influence. In a very real sense, the mind of each and every one of us is the battleground on which these influences meet, each warring for supremacy over the other. This is the spiritual reality of the struggle and effort that accompanies work-on-oneself. What must be accomplished cannot be achieved unaided. If we are to overcome the active forces of internal resistance and inertia, help of a special nature is required. We need vivification from God, as we do a relationship with Higher Life; this is the pivot around which we need to revolve.

This relationship marks the beginning of a partnership that will last for all eternity and is one of the prerequisites of work-on-oneself. We cannot achieve what we need to without Higher Help. Of ourselves we can do nothing but with God all things are possible. That is a truth I have come to know, understand, and love. None of us can work on ourselves unaided; we all need a higher power to bring about transformation, inner change, and spiritual growth. The desire to change, to grow and to transform ourselves has its origin in the depths of our being, and our will to do so constitutes our outreach to Higher Life, but the power to actualise those efforts has its source in God. The subsequent partnership between ourselves and God heralds a journey of self-discovery that will not only take us back to God, it will also take us through God, and it will enable us to live in God.

There is an enduring truth which states simply, *God is love*. The 15th century Indian poet and mystic saint Kabir (1440-1518) wrote that God, above all, is revealed in love. "One Love it is that pervades the whole world, few there are who know it fully" (Ferguson, 1976, p. 99). Until we personally experience this, until we *feel* this love flowing through our being and through the living universe, we can't understand the significance of those words. Irrespective of our vocation or external circumstances, all of us can have a rich and meaningful relationship with Higher Life if we want to. Although external forms of religious worship often inspire a sense of communion and companionship with others, love for God, *agape*, transcends any religious or societal act or expectation that might place us under its influence.

We find an echo of this is the writing of 17th century Church of England priest and English mystic William Law (1686-1761):

> Now there is but one possible way for a man to attain to salvation. There is not one for the Jew, another for the Christian and a third for the heathen. No! God is one, human nature is one, salvation is one and the way to it is one; and that is the desire of the soul turned towards God. When the desire is alive and breaks forth in any creature under heaven, then the lost sheep is found and the shepherd has it on his shoulder. (Ashby, 1828, p. 64).

The journey of the soul back to the source begins with a yearning heart and is born in the desire of the individual, any individual, whose mind is turned toward God. Religions are indeed many, but God is One. There are many paths back to the source; what does the name of each matter so long as the path brings us back to God?

As mentioned in the introductory pages of this book, I've incorporated written material published by Dr Groves, personal communications between Dr Groves and myself, information derived from my own hand-written notes taken in class lectures given by Dr Groves, selective autobiographical accounts and my understanding of the Work. Through all of this, through my ups and downs in life and in my difficulties and struggles with myself, I have always sought refuge and guidance in and from God. There have been moments of existential or spiritual crises and transformation, and although I worked with all the strength I could possibly muster under the circumstances, I always knew in my heart that the real working of the Work, and its vivification in my life, was carried out by God, the Divine, Yahweh, the Omnipotent, the Unnameable, and that the role I played in the unfolding drama was always one of collaboration.

> Beyond the space-time framework of the material world is the realm of creative causes. It operates with

> a fundamental benevolence and cares deeply for the sentient beings it brings forth. But since all of us have the freedom to exercise our initiative and to follow our personal drives, we often generate actions which interfere with other people. This leads to hostility, antagonism, strife and even war. Human beings create the evils of life by placing their own egotistic demands for satisfaction and power above the rights and freedoms of others. Belligerence, war, hatred, suspicion and jealousy are self-generated aspects of the sleeping human condition and were never created by Higher Life. (Groves, 2000, p. 11).

We are all prisoners of the sleep-state and the multitude of 'I's' that live within us. To further complicate matters, we live in an 'Adrenal Age' where the need and demand for new and ever-changing technology has become a driving force in human affairs, and while great and beneficial discoveries have been made in many fields, socially mankind has advanced very little. From an objective perspective it can be plainly stated that socially mankind is in a barbaric state, while spiritually he is in a desperate state. A great flood of negativity has entered the human psyche to such a degree that despite our great technological, medical, and scientific advances, spiritually we are living in the metaphorical Dark Ages.

> The earth is the seminary of heaven, and the human soul must grow up in a physical body in order to crystallize within itself an inner body that is immortal, since it is only on the earthly level that such a thing is possible. (Groves, 2011, p. 38).

Despite mankind's state, Groves reminds us that we all stand at the helm of our own eternal spiritual destiny. All of us can choose to embark on our Second Education so as to navigate our way into eternal life. Higher Life holds its breath, waiting, while God reaches toward us offering the opportunity to make something of ourselves, offering the invitation to participate in the unfolding of creation and the maintenance of the universe. Whether we accept the invitation or not is up to us.

Since I encountered and entered the Work much of my behaviour and many of my attitudes have changed. Initially it was as though heaven had opened its very gates to me and filled my life to overflowing. Then things began to go wrong, problems arose and difficulties that I could never possibly have imagined presented themselves and had to be overcome. The difficulties were surmounted, change and transformation occurred, and I began to wake up. And yet in understanding how far I've come; I also see how very far there is to go. If I were to die right now, would I be in a fit state to enter the spiritual universe? No, I would not.

I have written throughout that our relationship with God is the most important relationship of our life. For me, it *is* life. In my childhood I knew and felt with complete certainty and assuredness that something incredibly vast and loving stood behind the visible material world in which I lived. I knew that what my eyes showed me to be the world acted as a screen preventing me from seeing the real world which lay beyond it. Magnetic Centre led me to the Work, the Fourth Way and Beyond the Fourth Way, and the collective action of my engagement with each has enabled me to merge my way of being in the world with an intellectual understanding of the cosmological truths of the spirit and of multidimensional reality and to finally understand what I felt and knew as a child.

Esoteric knowledge provides a map of our existence and our spiritual destiny. It paves the way for the influx of Higher Life while simultaneously offering practical solutions to combat the trials and tribulations inevitably encountered while living a challenging planetary existence. It seems to me that it's only when we begin to feel the impact of the Work within our life that we begin to understand what the Work is, as we do its aim. Although we may be in the Work, it is nonetheless important to be flexible in our approach toward ourselves and life, and to be open to good and truth in all its forms. Nor can we hold to prejudices by which we not only judge others, but which generate unnecessary ill-feeling, ill-will and antagonism which the following life experience illustrates.

In January 2015 I presented a paper at a local conference in Sydney which showcased the afterlife. In introducing my presentation, I spoke a little about some of the influences in my life and briefly mentioned the esoteric School I had attended. After my presentation was finished, a woman spoke to me wanting to know more about the School and my experiences. After listening to my circumspect account, she then proceeded to tell me in an authoritative manner about her experiences in Konya, where she had been invited to go to London to become a member of a Sufi order (I too am a member of a Sufi order based in America, but I kept that to myself). The conversation came to an abrupt halt when she said with a condescending tone, "I was in an order you see, but you, you were in a cult" *(Authors journal notes, 26 January 2015).*

This exchange stifled any further conversation between us, and indeed we sat at the table in obvious silence. While it was apparent that the woman believed she had experienced a meaningful event in her life, what was also apparent was her need to categorise me in accordance with her understanding of the Work and its dissemination through Schools.

> The duty of every human being is to seek the truth, receive the truth, believe the truth, and live the truth. Truth is not the exclusive property of any single

> teacher, group, church or organisation. Truth is universal Divine substance which always leads the way to what is good, real and sacred. We can never receive into ourselves enough truth; it is infinite and endless, and always illuminates our lives with consciousness. We are greatly mistaken if we assume that any one revelation or communication of truth is going to tell us everything. It is our task to render ourselves into unprejudicial receptacles of truth, and to be able to love truth because it is truth. (Groves, 1998c, p. 7).

We are indeed greatly mistaken if we assume that any one revelation or communication of truth is going to tell us everything, as the experience which occurred at the conference illustrates. We simply cannot afford to be one-sided, or arrogant in our summation of knowledge because to do so risks limiting our understanding of truth. Furthermore, to do so creates unnecessary barriers between us and others.

> Far too many human spirits live in a state of enslavement to their prejudices and preferences; many spiritual communities are in a state of unnecessary mutual antagonism and it is impossible for them to realise their full potential. The celestials are not bound by prejudices and one-sided views; they are open to good and truth in all their forms, and rejoice when they see these things reflected, even dimly, in the lives of lesser beings. (Groves, 1998c, p. 6).

The cosmology of worlds within worlds is a science. Work-upon-oneself involves a lifetime of self-examination and self-cleansing as it does an active binding back to God, a re-membering. Our Second Education calls to us and it seems to me that our embodied lives are a purposeful means to an end in a future world very different to the one we call 'home'. If I ask myself what it is that I'm living for, what is my answer? I have to 'live' until I die, but it is *how* I choose to live, as it is what I am living *for* that is the primary consideration. What am I living for? The answer is so clear; I'm living to learn how to live.

I titled this chapter *A New Way*, because when we encounter the Work we do experience life and ourselves in a new way. A new beginning, or way, can take various forms, but what they all have in common is that they are start-points. They are that moment in time when our life, or aspects of our lives, change or move in a different direction, and this is precisely what occurs when we encounter the Work. We change because our thinking changes, which in turn changes our being, and because our being changes our spiritual destiny changes also.

In coming to the end of this chapter, I've chosen to include an extract of a dialogue, recorded verbatim between Ms Melissa Marston and Sri Govindananda Bharati (1826-1963) Hindu saint and otherwise known as the Shivapuri Baba. During 1961 and 1962 Ms Marston, her sister Marjorie and others accompanied John G. Bennett, Fourth Way teacher and author, on two visits to Sri Govindananda's Kathmandu retreat in Nepal. It is the final sentence in this dialogue which highlights the challenge that confronted me when Dr Groves asked me to put all of the Work into a book; I can experience the Work but I can never truly explain it, for that relationship can only exist and be understood by the individual and God.

M.M. There is always the feeling, perhaps more in the West than in the East, of not really understanding why we are here. That our life is passing, and somehow, we are not really using it as we might.

The Shivapuri Baba, who had listened carefully to all they told him, was sitting out of doors in an armchair holding a rose in his hand. We were sitting in front of him on low stools. He smiled, held out the rose and said:

S.B. Now – so to say – there is smell in this flower. Can I explain that smell so that you will smell it yourself?

M.M. No, I suppose not.

S.B. However I explain, you cannot know the smell. If I give it in your hand and you smell it, you know what the smell is. Is that not so? In the same way, I cannot answer your questions. Your questions will be answered only by God. See Him first. Every mystery is solved. Before that, whatever answer I can give you, it will not solve your problems. The first thing is to reach God. By seeing God, everything is known.

M.M. How can we do that?

S.B. Think of God alone. Dismiss every thought from your mind. You will see God.

M.M. That is very difficult.

S.B. It is difficult, but not impossible. It is difficult – I do know. But if you will take enough trouble to remove that difficulty, all other difficulties of life will vanish. Unless we see God, we cannot know anything. Before the sun comes, we can see nothing on the ground. When the sun comes, we see everything. So the Presence of God. When we know God we know everything. Before that, a simple explanation is no explanation at all. I can say, this is sweet smelling. But how many flowers there are that have a sweet smell! What is the nature of this smell? Unless you smell it, you cannot know. One can experience it, but it can never be explained. (Bennett, 1965, pp. 44-45).

AFTERWORD

A FINAL NOTE

In December 1998 Dr Groves gave his final lecture to students at his School; it was the last time I saw him in person. The room was filling fast, and I had left my seat to have a quick chat with a friend who was sitting toward the front of the lecture hall. When he entered, and before the class commenced, he approached each person, held their hand and spoke to them. Although my seat was further back in the room, I was unable to move with any decorum and before I knew it, he was standing in front of me.

He bent down and looked into my eyes, and as always, I felt his great kindness and warmth sweep over me. He held my small hand ever so gently in his large one, just like he did when I had first met him, and said, "All my very best to you, for your happiness, for your life. I give you blessings, blessings to you, blessings." Then he came even closer and said almost conspiratorially, and with a twinkle in his eye, "It may not be in this life that we will meet again."

Even then he knew my thoughts, so I simply nodded because I had been wondering if it might be possible to perhaps meet with him in person one more time. After he had spoken however, I wondered what it was that he knew that I didn't. He paused a moment and with a knowing smile said, "Death stalks us, and we never know when it will meet us." As he gently unclasped his hand from mine and straightened up, he said with a subtle joy and enthusiasm, "Live your life to the fullest." As he moved to the next person, I was able to make my way back to my seat. Suddenly he was in front of me again reaching for my hand. He smiled gently and once again his fingers closed over mine and warmth enveloped me. He spoke to me again, but the words were lost in what happened next.

The first time I had met him something odd had transpired, and somehow it seemed fitting that the last time I was to see him in person something other-worldly would also occur. As he held my hand in his, something passed from him into me, an energy of some sort. I felt it moving down his arm, through his hand and into mine, and then suddenly I saw a great white star, in the region of my solar plexus, burst into life. I felt the energy pulse before radiating outward from the solar plexus, slowly spreading throughout my body, permeating my being and saturating my cells. And yes, as on the occasion when I had first met him, he saw that I could see what was occurring, and he smiled gently at me in understanding, and that was that.

After Dr Groves died the School in its external sense closed and its students dispersed. Schools come and go, that is their nature and while some endure for tens, hundreds or thousands of years others live a short yet brilliant life, blazing across humanity like a shooting star across the night sky. Some students stayed in contact with one another, but I metaphorically wandered in the wilderness, and it was not until September 2014, some fifteen years later that I briefly encountered some fellow students. It was at this time also, that I decided to return to the manuscript with the view of finally publishing it. All was proceeding smoothly and then one May evening in 2015 I had an unexpected encounter. I had gone to bed and fallen asleep. During the night I woke suddenly to find myself standing in front of Dr Groves. I knew immediately that I wasn't dreaming but was in fact out of my body and in the spiritual universe. He was full of life and energised from within, and as he had always towered over me in life so too did he now; I had to tilt my head so I could look up into his face. He was smiling and pleased to see me, and I suddenly thought back to the last time that I had seen him, in December 1998.

I recalled his goodbyes to all present and remembered that when he approached me, I had been thinking whether it might be possible to see him again. At that moment he had taken my hand and knowing my thoughts had said, "It may not be in this life that we will meet again." These reflections were in my mind as I stood in front of him now, and as I looked into his face, I saw that his smile was a knowing one; I had the distinct impression that if he had verbally spoken he would have gleefully said, "Told you!" but he didn't need to verbalise the words because I already knew them.

With lightning speed and clarity of thought we understood one another instantly. The words he spoke to me during that last encounter were a prelude to the event he knew would take place in the future, to the event that was occurring now. And he was right. It wasn't in 'this' life embodied life that is, that we were to meet again, it was the next life, the afterlife. I suddenly jack-knifed up and found myself back in my body sitting up in bed in the dark, my mind racing. He had seen everything; the reconnecting with fellow students, the manuscript being published, and those books on shelves. And how fitting that this event should occur precisely when the manuscript was being edited and readied for future publication; it was a validation.

Although this *Final Note* is an addition to the text of the original manuscript, I feel the reader will understand why it's been included. Teachers of the Work and men and women of High Degree in all cultures and centuries leave their mark on humanity in various ways. Sometimes it's in the tales of their deeds told, sung, or written about by others. Sometimes it's in their art, their writing, their music or in the monuments they build, and sometimes it's because they have stood at the threshold of time where worlds meet. Those who have gone before us serve as doorways through which access to esoteric knowledge and spiritual instruction is made available to those who seek it. The enneagram of this manuscript is now complete, and since it came into being through the words of Dr Groves, I would like it to reside in his wisdom and to be cloaked with the following insight, "Real knowledge is the product of intuition, not of formal learning; intuition is that insight which comes from the very depths of being" (Groves, 1997c, p. 4).

ACKNOWLEDGMENTS AND PERMISSIONS

I acknowledge and particularly thank the efforts of Dr Philip W. Groves, who reviewed and approved the initial manuscript and who provided permission for me to quote freely from his published and non-published writings from the public and private domain. I also acknowledge and thank his permission for me to draw on our private correspondence, our private conversations and notes, my private journals, and my personal observations made during the 16 years I attended his School.

I acknowledge and thank Liz Coggins from Triam Press who reviewed and edited various versions of the manuscript and Robert Bryce from Studio Temenos, who reviewed and provided editing feedback on the final draft of the manuscript.

I gratefully acknowledge the following Permissions and Fair Dealings to reproduce copyright material:

Cover photo *Mi Serenata*, Robert Bryce of Studio Temenos, copyright © 2020 by Robert Bryce. Reprinted with permission of Robert Bryce, photographer.

Quotations from *The Unknowable Gurdjieff*, by Margaret Anderson, copyright © 1962 by Margaret Anderson. Published by The Penguin Group. Orphan work in the public domain, reprinted as Fair Use.

Quotations from *The Fiery Fountains*, by Margaret Anderson, copyright © 1969 by Margaret Anderson. Published by Horizon Press. Orphan work in the public domain, reprinted as Fair Use.

Quotations from *The Reality of Being: The Fourth Way of Gurdjieff*, by Jeanne de Salzmann, copyright © 2010 by the Heirs of Jeanne de Salzmann. Used by arrangement with The Permissions Company, LLC on behalf of Shambhala Publications Inc., Boulder, Colorado, www.shambhala.com

Quotations from *The Song Celestial: A poetic version of the Bhagavad Gita*, by Sir Edwin Arnold. Published by The Theosophical Publishing House. This material was reproduced by permission of Quest Books, the imprint of The Theosophical Publishing House, Wheaton Illinois.

Quotations from *Extracts from the letters and writings of William Law*, by Maria Ashby, copyright © 1828 by Maria Ashby. Published by Harvey and Darton. Orphan work in the public domain, reprinted as Fair Use.

Quotations from *The Crisis in Human Affairs*, by John G. Bennett, copyright © 1954. Published by Hodder & Stoughton Limited. Used by arrangement with the J G Bennett Foundation www. jgbennett.org

Quotations from *The Long Pilgrimage: The Life and Teaching of Sri Govindananda Bharati, known as the Shivapuri Baba*, by John G. Bennett, copyright © 1965. Published by Hodder & Stoughton Limited. Used by arrangement with the J G Bennett Foundation www.jgbennett.org

Quotations from *The Way to God: Selected Writings from Mahatma Gandhi*, edited by M.S. Deshpande, published by North Atlantic Books, copyright © 2009 by North Atlantic Books. Reprinted with permission of publisher.

Quotations from *Templeton Prize: For Progress Toward Research or Discoveries About Spiritual Realities: A Chronicle*, by Professor George

F. R. Ellis, copyright © 2004 by Professor George F. R. Ellis. John Templeton Foundation. Reprinted with permission of author.

Quotations from *The Tibetan Book of the Dead*, by Walter Y. Evans-Wentz, copyright © 1960 by Walter Y. Evans-Wentz. Published by Oxford University Press. Orphan work in the public domain, reprinted as Fair Use.

Quotations from *An Illustrated Encyclopaedia of Mysticism and the Mystery Religions*, by John Ferguson, © 2012. Reprinted with kind permission of Thames & Hudson Ltd, London.

Quotations from *Physicalism* in Journal of Parapsychology, by Richard S. Gebelein. Published by Rhine Research Centre, copyright © 2013 by Rhine Research Centre. Reprinted with permission of publisher.

Quotations from *Search Within* in Ancient Wisdom and Modern Science, edited by S. Groff. Published by State University of New York Press, copyright © 1984 by State University of New York Press. Reprinted with permission of publisher.

Quotations from *Introduction to Spiritual Psychology, Spiritual Psychology Volume 1*, by Dr Philip W. Groves, copyright © 1996 by Dr Philip W. Groves. Published by Triam Press. Reprinted with permission of author.

Quotations from *A Second Education, Spiritual Psychology Volume* 3, by Dr Philip W. Groves, copyright © 1996 by Dr Philip W. Groves. Published by Triam Press. Reprinted with permission of author.

Quotations from *Beyond the Fourth Way*, by Dr Philip W. Groves, copyright © 1996 by Dr Philip W. Groves. Published by Triam Press. Reprinted with permission of author.

Quotations from *The Anatomy of Essence (Part 1)*, by Dr Philip W. Groves, copyright © 1997 by Dr Philip W. Groves. Published by The Centre for Integrative Studies. Reprinted with permission of author.

Quotations from *Inversion and Reversal. 1977 Lecture Series*, by Dr Philip W. Groves, copyright © 1997 by Dr Philip W. Groves. Published by The Centre for Integrative Studies. Reprinted with permission of author.

Quotations from *Sufism: A Work Process. 1977 Lecture Series*, by Dr Philip W. Groves, copyright © 1997 by Dr Philip W. Groves. Published by The Centre for Integrative Studies. Reprinted with permission of author.

Quotations from *Awakening and Egeiro. 1977 Lecture Series*, by Dr Philip W. Groves, copyright © 1997 by Dr Philip W. Groves. Published by The Centre for Integrative Studies. Reprinted with permission of author.

Quotations from *Order, Beauty and Meaning. 1977 Lecture Series*, by Dr Philip W. Groves, copyright © 1997 by Dr Philip W. Groves. Published by The Centre for Integrative Studies. Reprinted with permission of author.

Quotations from *Some Aspects of Being. 1998 Lecture Series*, by Dr Phillip W. Groves, copyright © 1998 by Dr Philip W. Groves. Published by The Centre for Integrative Studies. Reprinted with permission of author.

Quotations from *Knowledge and Being: A Stern Reminder. 1998 Lecture Series*, by Dr Philip W. Groves, copyright © 1998 by Dr Philip W. Groves. Published by The Centre for Integrative Studies. Reprinted with permission of author.

Quotations from *Hidden Aspects of Eastern & Western Christianity. 1998 Lecture Series*, by Dr Philip W. Groves, copyright © 1998 by Dr Philip W. Groves. Published by The Centre for Integrative Studies. Reprinted with permission of author.

Quotations from *The Organics of Life. 1998 Lecture Series*, by Dr Philip W. Groves, copyright © 1998 by Dr Philip W. Groves. Published by The Centre for Integrative Studies. Reprinted with permission of author.

Quotations from *Aspects of the Creative Process. 1998 Lecture Series*, by Dr Philip W. Groves, copyright © 1998 by Dr Philip W. Groves. Published by The Philip W. Groves Centre. Reprinted with permission of author.

Quotations from *From End to Beginning. 1998 Lecture Series*, by Dr Philip W. Groves, copyright © 1998 by Dr Philip W. Groves. Published by The Philip W. Groves Centre. Reprinted with permission of author.

Quotations from *Science & Esoteric Wisdom*, by Dr Philip W. Groves, copyright © 2000 by Dr Philip W. Groves. Published by Triam Press. Reprinted with permission of author.

Quotations from *Journey through the World of Symbol*, by Dr Philip W. Groves, copyright © 2011 by Dr Philip W. Groves. Published by Triam Press. Reprinted with permission of author.

Quotations from *Spiritual and Material Connections: The Esotericism of Emanuel Swedenborg, A Collection of Lectures*, by Dr Philip W. Groves, copyright © 2014 by Dr Philip W. Groves. Published by Triam Press. Reprinted with permission of author.

Quotations from *Life is Real Only Then, When 'I Am'*, by George I. Gurdjieff, copyright © by George I. Gurdjieff. Published by Viking Arkana. Reprinted with permission of publisher.

Quotations from *Australian Concise Oxford Dictionary*, edited by J. M. Hughes, P. A. Michell & W. S. Ramson. Published by Oxford University Press, copyright © 1992 by Oxford University Press. Reprinted with permission of publisher.

Quotations from *Undiscovered Country: A Spiritual Adventure*, by Kathryn Hulme, copyright © 1966 by Kathryn Hulme. Published by Little, Brown & Company Limited. Orphan work in the public domain, reprinted as Fair Use.

Quotations from *THE WORKS OF WILLIAM JAMES – THE PRINCIPLES OF PSYCHOLOGY: VOLUMES I–II*, Frederick Burkhardt, General Editor, Fredson Bowers, Textual Editor, Ignas K. Skrupskelis, Associate Editor, Cambridge, Mass.: Harvard University Press, Copyright © 1981 by the President and Fellows of Harvard College. Used by permission. All rights reserved.

Quotations from *Physics and Philosophy*, by Sir James H. Jeans. Published by Dover Publications, Inc. Orphan work in the public domain, reprinted as Fair Use.

Quotations from *Reductionism: Analysis and the Fullness of Reality*, by Richard H. Jones. Published by Associated University Presses, copyright © 2000 by Associated University Presses. Reprinted with permission of publisher.

Quotations from *Gandhi: A Spiritual Journey*, by Madhav V. Kamath, published by Indus Source Books, copyright © 2007 By Madhav V. Kamath. Reprinted with permission of publisher.

Quotations from *Ways of Being: The alchemy of bereavement and communique*, by Michele T. Knight, copyright © 2013 by Michele T. Knight. Reprinted with permission of author.

Quotations from *The Gospel of Thomas* (11,2) in The Nag Hammadi Library: The definitive new translation of the Gnostic scriptures, complete in one volume, by Helmut Koester & Thomas O. Lambdin. Published by HarperSanFrancisco & Brill, copyright © 1978 by HarperSanFrancisco & Brill. Reprinted with permission of HarperSanFrancisco & Brill.

Quotations from *The Gospel of Thomas*, by Jean-Yves Leloup. Published by Inner Traditions International and Bear & Company, copyright © 2005 by Inner Traditions International and Bear & Company. All rights reserved. http://www.Innertraditions.com Reprinted with permission of publisher.

Quotations from *Deification in Christ: The Nature of the Human*

Person, by Panayiotis Nellas, published by St Vladimir's Seminary Press, copyright © 1987 by St Vladimir's Seminary Press. Reprinted with permission of publisher.

Footnote reference to the book *Principles of Neurotheology*, by Professor Andrew Newberg, copyright © 2010 by Professor Andrew Newberg. Reprinted with permission of author.

Quotations from *Psychological Commentaries on the Teaching of Gurdjieff & Ouspensky. Volume 1*, by Maurice Nicoll, published by SHAMBHALA PUBLICATIONS, INC., copyright © 1980 by Isobel Salole. Reprinted with permission of Eureka Editions.

Quotations from *Psychological Commentaries on the Teaching of Gurdjieff & Ouspensky. Volume 4*, by Maurice Nicoll, published by SAMUEL WEISER, INC., copyright © 1980 by Isobel Salole. Reprinted with permission of Eureka Editions.

Quotations from *Psychological Commentaries on the Teaching of Gurdjieff & Ouspensky. Volume 5*, by Maurice Nicoll, published by Samuel Weiser, Inc., copyright © 1975 by Isobel Salole. Reprinted with permission of Eureka Editions.

Quotations from *The New Man: An Interpretation of Some Parables and Miracles of Christ*, by Maurice Nicoll, published by SHAMBHALA PUBLICATIONS, INC., copyright © 1981 by Isobel Salole. Reprinted with permission of Eureka Editions.

Footnote reference to the book *The Work Life*, by Beryl Pogson, published by SAMUEL WEISER, INC., copyright © 1994 by Lewis Creed. Reprinted with permission of Eureka Editions.

Quotations from *The Third Eye*, by Lobsang Rampa. Published by Secker & Warburg. Orphan work in the public domain, reprinted as Fair Use.

Quotations from *States of Grace*, by Charlene Spretnak, published by HarperSanFrancisco. Copyright © 1991 by Charlene Spretnak. Reprinted with permission of Frances Goldin Literary Agency.

Quotations from *The Concise Dictionary of English Etymology*, by Walter W. Skeat, published by Wordsworth Editions Ltd., copyright © 1993 by Wordsworth Editions Ltd. Reprinted with permission of publisher.

Quotations from *Arcana Coelestia, the Heavenly Arcana Unfolded, Vol V*, by Emanuel Swedenborg, published by SWEDENBORG FOUNDATION, copyright © 1984 by SWEDENBORG FOUNDATION. Reprinted with permission of publisher.

Quotations from *Heaven and its Wonders and Hell From Things Heard and Seen*, by Emanuel Swedenborg, published by SWEDENBORG SOCIETY, (Inc.), copyright © by SWEDENBORG SOCIETY, (Inc.). Orphan work in the public domain, Reprinted with permission of publisher.

Quotations from *De Miraculis*, by Emanuel Swedenborg, SWEDENBORG SOCIETY (Inc)., copyright © by SWEDENBORG SOCIETY, (Inc.). Orphan work in the public domain, Reprinted with permission of publisher.

Quotations from *A Brief Exposition of the Doctrine of The New Church signified by the New Jerusalem in The Revelation*, by Emanuel Swedenborg, SWEDENBORG SOCIETY, Incorporated, copyright © by SWEDENBORG SOCIETY Incorporated. Orphan work in the public domain, Reprinted with permission of publisher.

Quotations from *The Canons of the New Church or The Whole Theology of the New Church*, by Emanuel Swedenborg, THE SWEDENBORG SOCIETY (Inc)., copyright © by THE SWEDENBORG SOCIETY, (Inc.). Reprinted with permission of publisher.

Quotations from *Four Doctrines of the New Jerusalem*, by Emanuel Swedenborg, THE SWEDENBORG SOCIETY, copyright © by THE SWEDENBORG SOCIETY. Reprinted with permission of publisher.

Quotations from *The Spiritual Diary of Emanuel Swedenborg*, by

Emanuel Swedenborg, THE SWEDENBORG SOCIETY, copyright © by THE SWEDENBORG SOCIETY. Reprinted with permission of publisher.

Quotations from *The New Jerusalem and its Heavenly Doctrine according to what has been heard from Heaven with an Introduction concerning The New Heaven and the New Earth*, by Emanuel Swedenborg, THE SWEDENBORG SOCIETY, copyright © by THE SWEDENBORG SOCIETY. Reprinted with permission of publisher.

Quotations from *The passion of the western mind: Understanding the Ideas That Have Shaped Our World View*, by Richard Tarnas, published by Ballantine Books, copyright © 1991 by Richard Tarnas. Reprinted with permission of publisher.

Quotations from *The Pursuit of God*, by Aiden Wilson Tozer, copyright © 2015 by Aiden Wilson. Published by Aneko Press. Reprinted with permission of publisher.

Quotations from Mysticism: *The Development of Humankind's Spiritual Consciousness*, by Evelyn Underhill, copyright © 1995 by Evelyn Underhill. Orphan work in the public domain, reprinted as Fair Use.

Quotations from *Discovering Transpersonal Identity* in Journal of Humanistic Psychology, by Frances E. Vaughan. Published by Sage Publications, copyright © 1985 by Sage Publications. Reprinted with permission of publisher.

Quotations from *No Other Gods: Christian Belief in Dialogue with Buddhism, Hinduism, and Islam*, by Hendrik M. Vroom. Published by Wun. B. Erdmens Publishing Co., copyright © 1985 by Wun. B. Erdmens Publishing Co. Reprinted with permission of publisher.

Quotations from *Ode: Intimations of Immortality From Recollections of Early Childhood*, by William Wordsworth, copyright © 1884 by William Wordsworth. Orphan work in the public domain, reprinted as Fair Use.

Requests for permission to quote from the following sources have gone unanswered or are currently pending:

A quotation from the book *The Turning Point: Science, Society, and the Rising Culture*, by Fritjof Capra, published by Flamingo.

A quotation from the book *The Thirteen Principal Upanishads*, Second Edition, by Robert E. Hume (Trans.), published by Oxford University Press.

A quotation from the book *Charles Kingsley: His Letters, and Memories of His Life*, by Frances E. Grenfell (Ed.), published by Cambridge University Press.

A quotation from the book *In Search of the Miraculous: Fragments of an unknown teaching*, by Pyotr Ouspensky, published by Harcourt Brace Jovanovich.

A quotation from the book *A Sacred Place to Dwell: Living with reverence upon the earth*, by Henryk Skolimowski, published by Vega.

A quotation from the book *One Hundred Poems of Kabir*, by Rabindranath Tagore (Trans.), published by Macmillan and Co., Limited.

While every effort has been made to trace the copyright holders and obtain permissions, this has not been possible in all cases. Any omissions brought to our notice will be remedied and suitable action taken.

REFERENCES

Anderson, M. (1962). *The Unknowable Gurdjieff*. England: The Penguin Group.

Anderson, M. (1969). *The Fiery Fountains*. New York: Horizon Press.

Arnold, Sir E. (Trans.) (1975). *The Song Celestial: A poetic version of the Bhagavad Gita*. Quest Book Miniature. United States of America: The Theosophical Publishing House. (Original work published 1885).

Ashby, M. (1828). *Extracts from the letters and writings of William Law*. London: Harvey and Darton.

Bennett, J. G. (1954). *The Crisis in Human Affairs*. London, England: Hodder & Stoughton Limited.

Bennett, J. G. (1965). *Long Pilgrimage: The Life and Teaching of Sri Govindananda Bharati, known as the Shivapuri Baba*. London, England: Hodder & Stoughton Limited.

Capra, F. (1983). *The Turning Point: Science, Society and the Rising Culture*. London, England: Flamingo.

De Salzmann, J. M. (2010). *The Reality of Being: The Fourth Way of Gurdjieff*. Boston, Massachusetts: Shambhala Publications, Inc.

Deshpande, M. S. (Ed.) (2009). *The Way to God: Selected Writings of Mahatma Gandhi*. Berkley, California: North Atlantic Books.

Ellis, G. F. R. (2004). *2004 Templeton Prize: For Progress Toward Research or Discoveries About Spiritual Realities: A Chronicle*. John Templeton Foundation: Author.

Evans-Wentz, W. Y. (Ed.) (1957). *The Tibetan Book of the Dead* (3rd ed.). London, England: Oxford University Press.

Ferguson, J. (1976). *An Illustrated Encyclopaedia of Mysticism and the Mystery Religions.* London, England: Thames & Hudson.

Gebelein, R. S. (2013). Physicalism. *Journal of Parapsychology, 77*(2), 159-164.

Groves, P. W. (1996a). Introduction to Spiritual Psychology. In *Spiritual Psychology Volume 1.* Castlecrag, Australia: The Triam Press.

Groves, P. W. (1996b). A Second Education. In *Spiritual Psychology, Volume 3.* Castlecrag, Australia: The Triam Press.

Groves, P. W. (1996c). *Beyond the Fourth Way.* Castlecrag, Australia: The Triam Press.

Groves, P. W. (1997a). The Anatomy of Essence (Part 1). *1997 Lecture Series, 7*(13), 1-12. Balgowlah, Australia: The Centre for Integrative Studies.

Groves, P. W. (1997b). Inversion and Reversal. *1997 Lecture Series, 13*(26), 11-20. Balgowlah, Australia: The Centre for Integrative Studies.

Groves, P. W. (1997c). Sufism – A Work Process. *1997 Lecture Series, 17*(33), 1-11. Balgowlah, Australia: The Centre for Integrative Studies.

Groves, P. W. (1997d). Awakening and Egeiro. *1997 Lecture Series, 19*(38), 13-23. Balgowlah, Australia: The Centre for Integrative Studies.

Groves, P. W. (1997e). Order, Beauty and Meaning. *1997 Lecture Series, 23*(46), 15-25. Balgowlah, Australia: The Centre for Integrative Studies.

Groves, P. W. (1998a). Some Aspects of Being. *1998 Lecture Series, 2*(4), 11-21. Balgowlah, Australia: The Centre for Integrative Studies.

Groves, P. W. (1998b). Knowledge and Being: A Stern Reminder. *1998 Lecture Series, 3*(6), 11-20. Balgowlah, Australia: The Centre for Integrative Studies.

Groves, P. W. (1998c). Hidden Aspects of Eastern & Western Christianity. *1998 Lecture Series, 7*(13), 1-9. Balgowlah, Australia: The Centre for Integrative Studies.

Groves, P. W. (1998d). The Organics of Life. *1998 Lecture Series, 8*(15), 1-9. Balgowlah, Australia: The Centre for Integrative Studies.

Groves, P. W. (1998e). Aspects of the Creative Process. *1998 Lecture Series, 20*(40), 11-19. Balgowlah, Australia: The Philip W. Groves Centre.

Groves, P. W. (1998f). From End to Beginning. *1998 Lecture Series, 24*(48), 11-18. Balgowlah, Australia: The Philip W. Groves Centre.

Groves, P. W. (2000). *Science & Esoteric Wisdom*. Castlecrag, Australia: The Triam Press.

Groves, P. W. (2011). *Journey through the World of Symbol*. Castlecrag, Australia: Triam Press.

Groves, P. W. (2014). *Spiritual and Material Connections: The Esotericism of Emanuel Swedenborg, A Collection of Lectures.* Castlecrag, Australia: The Triam Press.

Gurdjieff, G. I. (1999). *Beelzebub's Tales to His Grandson: An objectively impartial criticism of the life of man*. United States of America: Penguin Arkana.

Gurdjieff, G. I. (1991). *Life is Real Only Then, When 'I Am'.* All and Everything Third Series. United States of America: Viking Arkana.

Hughes, J. M., Michell, P.A., Ramson, W.S. (Eds.) (1992). *Australian Concise Oxford Dictionary.* South Melbourne, Australia: Oxford University Press.

Hulme, K. (1966). *Undiscovered Country: A Spiritual Adventure*. Canada: Little, Brown & Company Limited.

Hume, R. E. (Trans.) (1931). *The Thirteen Principal Upanishads,* Second Edition, Revised. Oxford, England: Oxford University Press. (Original work published 1921).

Jain, Y. (1984). Search Within. In S. Grof (Ed.), *Ancient Wisdom and Modern Science* (pp. 110-114). Albany, United States of America: State University of New York Press.

James, W. (1950). *The Principles of Psychology, (Vol 1)*, First Edition, Revised. New York, America: Dover Publications. (Original work published 1890).

Jeans, J. H. (1981). *Physics and Philosophy*. New York, United States of America: Dover Publications, Inc.

Jones, R. H. (2000). *Reductionism: Analysis and the Fullness of Reality*. Cranbury, NJ: Associated University Presses.

Jung, C. G. (1995). *Memories, Dreams, Reflections.* London, England: Fontana Press.

Kamath, M. V. (2007). *Gandhi: A Spiritual Journey*. Mumbai, India: Indus Source Books.

Kingsley, C. (2011). *Charles Kingsley: His Letters, and Memories of His Life* (F. E. Kingsley, Ed.). Cambridge, England: Cambridge University Press. (Original work published 1877).

Knight, M. T. (2013). *Ways of Being: The alchemy of bereavement and communique* (Doctoral dissertation), University of Sydney, Australia: University of Sydney. Available http://hdl.handle. net/2123/13764

Koester, H. & Lambdin, T. O. (Trans.) (1978). The Gospel of Thomas (11,2). *The Nag Hammadi Library: The definitive new translation of the Gnostic scriptures, complete in one volume*, Revised Edition, pp. 124-138. United States of America: HarperSanFrancisco.

Needleman, J. (2005). *Foreword, The Gospel of Thomas*. Rochester,Vermont: Inner Traditions International.

Nellas, P. (1987). *Deification in Christ: The Nature of the Human Person.* Crestwood, N.Y., United States of America: St Vladimir's Seminary Press.

Newberg, A. (2010). *Principles of Neurotheology*. Surry, England: Ashgate Publishing Limited.

Nicoll, M. (1984). *The New Man: An Interpretation of Some Parables and Miracles of Christ.* Colorado, United States of America: SHAMBHALA PUBLICATIONS, INC.

Nicoll, M. (1985). *Psychological Commentaries on the Teaching of Gurdjieff & Ouspensky*. Volume 1. Boston, Massachusetts: SHAMBHALA PUBLICATIONS, INC.

Nicoll, M. (1996). *Psychological Commentaries on the Teaching of Gurdjieff & Ouspensky*. Volume 4. York Beach, Maine: SAMUEL WEISER, INC.

Nicoll, M. (1996). *Psychological Commentaries on the Teaching of Gurdjieff & Ouspensky*. Volume 5. York Beach, Maine: Samuel Weiser, Inc.

Nicoll, M. (1998). *Simple Explanation of Work Ideas.* Utrecht, The Netherlands: Eureka Editions.

Ouspensky, P. D. (1949). *In Search of the Miraculous: Fragments of an unknown teaching*. United States of America: Harcourt Brace Jovanovich.

Pogson, B. (1994). *The Work Life.* York Beach, Maine: SAMUEL WEISER, INC.

Rampa, L. (1956). *The Third Eye*. United Kingdom: Secker & Warburg.

Skeat, W. W. (1993). *The Concise Dictionary of English Etymology.* Hertfordshire, United Kingdom: Wordsworth Editions Ltd.

Skolimowski, H. (2002). *A Sacred Place to Dwell: Living with reverence upon the earth*. Great Britain: Vega.

Spretnak, C. (1993). *States of Grace: The recovery of meaning in the postmodern age.* New York: HarperSanFrancisco.

Swedenborg, E. (1937). *Heaven and its Wonders and Hell From Things Heard and Seen* (Rev. J. R. Rendell, Trans.). London, England: SWEDENBORG SOCIETY, (Inc.) (Original work published 1758).

Swedenborg, E. (1947). *De Miraculis* (Rev. P. H. Johnson, Trans.). London, England: SWEDENBORG SOCIETY (Inc). (First edition published posthumously 1943).

Swedenborg, E. (1952). *A Brief Exposition of the Doctrine of The New Church signified by the New Jerusalem in The Revelation* (R. Stanley, Trans.). London, England: SWEDENBORG SOCIETY, Incorporated. (Original work published 1769).

Swedenborg, E. (1954). *The Canons of the New Church or The Whole Theology of the New Church* (F. F. Coulson, Ed.). London, England: THE SWEDENBORG SOCIETY (Inc). (Original work written 1769, translated and published posthumously).

Swedenborg, E. (1967). *Four Doctrines of the New Jerusalem* (Wm. C. Dick,

Trans.). London, England: THE SWEDENBORG SOCIETY. (Original work published 1763).

Swedenborg, E. (1977). *The Spiritual Diary of Emanuel Swedenborg*. London, England: THE SWEDENBORG SOCIETY. (Published posthumously, 1883-1902).

Swedenborg, E. (1984). *Arcana Coelestia, the Heavenly Arcana Unfolded, Vol V* (Rev. J. Clowes, Trans.). New York, United States of America: SWEDENBORG FOUNDATION. (Original work published 1751-1752).

Swedenborg, E. (1993). *The New Jerusalem and its Heavenly Doctrine according to what has been heard from Heaven with an Introduction concerning The New Heaven and the New Earth* (Rev. R. L. Tafel, Trans.). London, England: THE SWEDENBORG SOCIETY. (Original work published 1758).

Tagore, R. (Trans.) (1915). One Hundred Poems of Kabir. London: Macmillan and Co., Limited.

Tarnas, R. (1991). *The passion of the western mind: Understanding the Ideas that Have Shaped Our World View*. New York, America: Ballantine Books.

Tozer, A. W. (2015). *The Pursuit of God*. Abbotsford, America: Aneko Press.

Underhill, E. (1995). *Mysticism: The Development of Humankind's Spiritual Consciousness.* London, England: Bracken Books.

Vaughan, F. E. (1985). Discovering Transpersonal Identity. *Journal of Humanistic Psychology, 25*(3), 13-38.

Vroom, H. M. (1996). *No Other Gods: Christian Belief in Dialogue with Buddhism, Hinduism, and Islam.* Grand Rapids, Michigan: Wun. B. Erdmens Publishing.

Wordsworth, W. (1884). *Ode: Intimations of Immortality From Recollections of Early Childhood.* Boston, United States of America: D. LOTHROP AND COMPANY.

Made in the USA
Monee, IL
04 November 2026

67378327R00125